PENGUIN BOOKS

THE RESTLESS EARTH

Born in London in 1931, Nigel Calder was educated at the Merchant Taylors' School and at Cambridge, where he earned a master's degree in natural sciences. He served his apprenticeship as a science writer on the staff of *New Scientist* and from 1962 to 1966 was its editor. Winner in 1972 of UNESCO's Kalinga Prize for the Popularization of Science, he has traveled around the world five times, seeking out the latest scientific knowledge and interpreting it for his readers. Books by him include *The Weather Machine, The Human Conspiracy, The Life Game, The Mind of Man, Violent Universe, The Key to the Universe,* and *Spaceships of the Mind*; and he has contributed numerous articles to both British and American publications, among the latter *Science, Science Year,* and the *Nation.* A devoted sailor, Calder lives with his wife and five children in Sussex, near the coast.

Nigel Calder

The Restless Earth

A Report on the New Geology

PENGUIN BOOKS

Penguin Books Ltd, Harmondsworth,
Middlesex, England
Penguin Books, 625 Madison Avenue,
New York, New York 10022, U.S.A.
Penguin Books Australia Ltd, Ringwood,
Victoria, Australia
Penguin Books Canada Limited, 2801 John Street,
Markham, Ontario, Canada L3R 1B4
Penguin Books (N.Z.) Ltd, 182–190 Wairau Road,
Auckland 10, New Zealand

First published in Great Britain
by the British Broadcasting Corporation 1972
First published in the United States of America
by The Viking Press 1972
Viking Compass Edition published 1973
Reprinted 1974, 1975, 1977
Published in Penguin Books 1978
Reprinted 1979

Copyright © Nigel Calder, 1972
All rights reserved

ISBN 0 14 00.4902 9

Text printed in the United States of America
by The Book Press, Brattleboro, Vermont
Set in Palatino

Contents

Acknowledgments

Acknowledgment is due to the following for their permission to reproduce illustrations in this book

Aerofilms, 69, **104**; Alpha, **16**; Ambraseys, N. N., 29, 30; American Museum of Natural History, 131; Barbetti, M. F., and McElhinny, M. W., 38; Barghoorn, E. S., and Schopf, J. W., 116 (2 and 3); Black Star (J. Byerman), 59; British Petroleum Company Ltd, 78; British Steel Corporation, **137**; Brown Brothers, 84 (top); California Institute of Technology, 93; Camera Press, (A. Gregory) 20 (bottom), 101, 109, (Almasy) 145; Canadian Government Air Photo Division, 76; Cash, J. Allen, 52; Daily Telegraph, **32**; De Wys Inc., (Everett C. Johnson) 84 (bottom), (Svat Macha) 87; Flemming, N. C., 17; Geological Survey of Canada (J. A. Donaldson), 118; Geological Survey of Greenland, 119; Glaessner, M. F., 120 (bottom); Hale Observatories, 129; Hamersley Iron Pty Ltd, 132; Icelandic Photo and Press Service, 53, 55, 102; Institut Français du Petrole, 67; Institute of Geological Sciences, 10, 66, **81**; Kennedy, W. J., 115; Khan, M. A., 117; Le Pichon, X., Scientific Group, Centre Océanologique de Bretagne, 14; Luce, J. V., **17**; Muench, J., 61; Nagy, B., and L. A., 116 (1); NASA, **49**, 57, 94, 95; Photo Library Inc., 139 (bottom), 140; Photo Researchers Inc., 83; Ramberg, H., 46; Rapho Guillemette, (Ned Haines) 88; Rinehard, S., 135; Roberts, G. R., 141; Röttger, R., Geologisches Institut, Kiel, **121**; Scott Polar Research Institute, Cambridge, 47; Scripps Institution of Oceanography, 40; Shelton, J. S., 20 (top), 36, 74, 123, 139 (top); Smith, A. G. from *Understanding the Earth* (Artemis Press), 23; Smith, Edwin, 110, 149; Sturmer, W., 120 (top); Swiss National Tourist Office, 27; Talbot, R. W., 103; Trans-Antarctic Expedition, 114; University of East Anglia, 113; Vine, F. J., **65**; Washburn, Bradford, 126; Wegener, Alfred, *The Origin of Continents and Oceans* (Methuen), 43 (2); Wyckoff, J., 68; ZFA, **48**, **80**, **105**, **120**, **136**.

Figures in **bold** indicate colour plates facing.

Maps and diagrams by Diagram.

Author's note

This book is one outcome of world-wide travels undertaken during the preparation of a major television programme that has been produced by Philip Daly and scripted by the author. The book enlarges on material gathered for the programme, although it has been separately conceived.

For the opportunity to make the journey, acknowledgments are due to the BBC and to its overseas partners in the television production, namely National Educational Television (USA), Sveriges Radio, the Australian Broadcasting Commission and Norddeutscher Rundfunk.

Some 200 earth scientists in a dozen countries gave generously of their time and information; to them, my warmest thanks. A book that ranges widely cannot deal adequately with the work of any of these individuals; indeed, in many instances my informants have had to remain unnamed. These apparent discourtesies all round will perhaps be forgiven if I have succeeded in passing on something of the corporate excitement and intellectual achievement of the present revolution in the earth sciences, which were so amply communicated to me. I am grateful to John Sutton of Imperial College, London, for encouraging a certain audacity in my approach to difficult issues.

Further reading – Because so much of the story is new, many books on geology are seriously out of date. On the other hand, I have dealt briefly with subjects that are well worth going into in more detail and the following books are recommended:

Debate about the Earth (geophysics and continental drift) by H. Takeuchi, S. Uyeda and H. Kanamori (Freeman, Cooper; San Francisco 1970)

Physics of the Earth (a simple introduction to geophysics) by T. F. Gaskell (Thames and Hudson; London 1970)

The New World of the Oceans ('men and oceanography') by Daniel Behrman (Little, Brown; Boston 1969)

The Origin of Oceans and Continents (the manifesto of continental drift) by Alfred Wegener (republished Methuen; London 1967)

The Earth We Live On (history of geology) by Ruth Moore (Knopf; New York 1971)

Principles of Physical Geology (still the finest textbook) by Arthur Holmes (Nelson; London 1965)

Understanding the Earth (a reader for undergraduates) edited by I. G. Gass, Peter J. Smith and R. C. L. Wilson for the Open University (Artemis; Sussex 1971)

The television programme *The Restless Earth* was broadcast on BBC2 on 16 February 1972. The executive producer was Philip Daly, assisted by Colin Riach and Lars Wallén. Gene Carr was the principal film cameraman and Roger Waugh the film editor. Graphics were by Charles McGhie and John Horton. The studio designer was Stewart Marshall and the studio director was Richard Finny. The programme was written by Nigel Calder.

The Restless Earth

Rock layers that formed flat on the sea floor, during many millions of years, were here upended in a collision between North America and Europe. Then, in turn, new rock layers formed flat over them, in the course of further millions of years. After the early geologist James Hutton had taken John Playfair to this site at Siccar Point, Scotland, Playfair wrote: 'The mind seemed to grow giddy by looking so far into the abyss of time.'

A sense of time

Ten years is a long time in modern science and a long time in the life of a man. But to our planet ten years is almost nothing. It is scarcely long enough to add a tenth of an inch to the great thicknesses of rock that grow by the accumulation of mud on the bed of a shallow sea. It allows time for a few volcanoes to erupt and for the release of a multitude of earthquakes, but not for the continents to shift their ground by much more than a foot or so. Only a sharp-eyed man will notice the Earth change in his lifetime, except in its most active zones.

In the company of Philip Daly of the BBC, I have visited some of those active zones. With guidance from the earth scientists, we found evidence of the planet's restlessness in Africa, Japan, California and elsewhere. The processes that reshape the Earth are now easy enough to grasp and are often verifiable by eye. What taxes thought is the rate at which these processes operate.

Given time, hard rock flows like pitch. Given time, the continent where you sit could split open and allow a new ocean basin to form, as broad as the Atlantic. Or another continent might bear down upon your favourite beach and heave it above the snowline. Such events are commonplace in geological maps of the new style, but nothing of the kind will happen summarily in our lifetimes. The Earth takes about 100 million years, a megacentury, to engineer or dismantle a major ocean basin.

By scrutiny of fossils and of subtleties in the atomic composition of rocks (see p. 91) geologists can put dates to events in the Earth's history with fair confidence. In the chapters that follow several very different calendars unavoidably coexist. The familiar *anni dominorum* mark the progress in human comprehension of the Earth, during recent centuries and especially during the past decade. Another calendar takes us back thousands of years, to important geophysical events occurring in human prehistory. The time-scale is again stretched greatly for rearrangements of the face of the globe occupying tens or hundreds of millions of years. Finally, there is the whole span of the Earth's existence, reckoned in thousands of millions of years.

Making light of this great burden of time, I use the abbreviation —500 MY (for example) to mean 500 million years ago. But the origin of the Earth and the birth of life upon it are, in a sense, as far away from us in time as the faintest galaxies are from us in space, in the distant reaches of the universe. A million years is a figure as meaningless as a million light-years to a mind that lives for only an ultramicroscopic fraction of that time. Like the speed of light or the computed size of an atom, it is a quantity for the intellect to reason with rather than for the senses to grasp. With the age of the Earth estimated at 4600 million years, we can make only very inadequate comparisons with familiar things.

For instance, the Earth is so old that the huge continents of the planet have been amassed at an average rate of only ten acres a year. Had Atlas been paid a penny a day for holding the Earth for all this time, he could have bought up Fort Knox long ago.

Or we can depict Mother Earth as a lady of 46, if her 'years' are megacenturies. The first seven of those years are wholly lost to the biographer, but the deeds of her later childhood are to be seen in old rocks in Greenland and South Africa. Like the human memory, the surface of our planet distorts the record, emphasising more recent events and letting the rest pass into vagueness – or at least into unimpressive joints in worn-down mountain chains.

Most of what we recognise on Earth, including all substantial animal life, is the product of the past six years of the lady's life. She flowered, literally, in her middle age. Her continents were quite bare of life until she was getting on for 42 and flowering plants did not

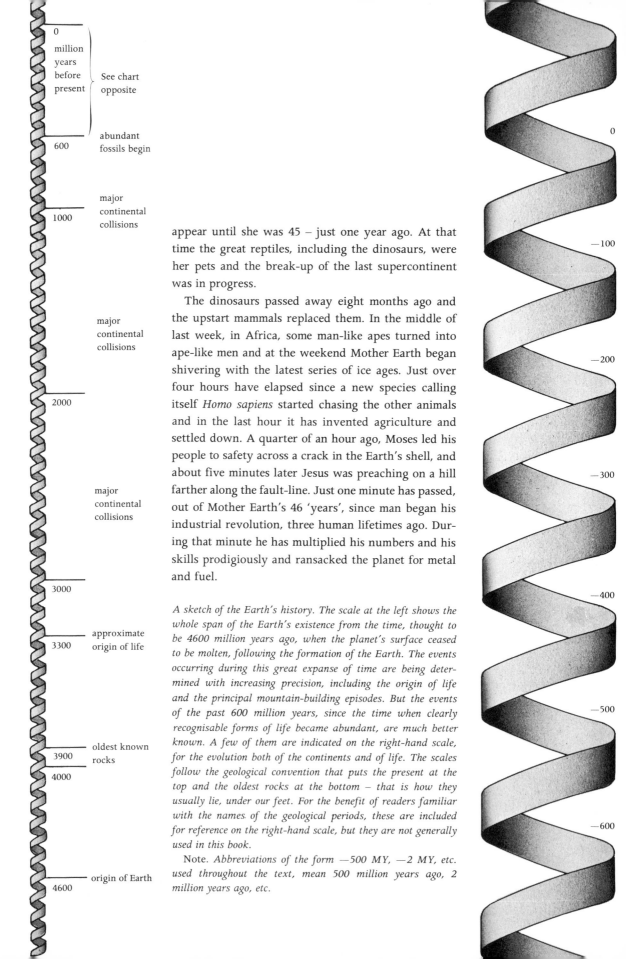

0
million
years
before
present
} See chart
opposite

600 — abundant fossils begin

1000 — major continental collisions

2000 — major continental collisions

3000

3300 — approximate origin of life

3900 — oldest known rocks

4000

4600 — origin of Earth

0
−100
−200
−300
−400
−500
−600

appear until she was 45 – just one year ago. At that time the great reptiles, including the dinosaurs, were her pets and the break-up of the last supercontinent was in progress.

The dinosaurs passed away eight months ago and the upstart mammals replaced them. In the middle of last week, in Africa, some man-like apes turned into ape-like men and at the weekend Mother Earth began shivering with the latest series of ice ages. Just over four hours have elapsed since a new species calling itself *Homo sapiens* started chasing the other animals and in the last hour it has invented agriculture and settled down. A quarter of an hour ago, Moses led his people to safety across a crack in the Earth's shell, and about five minutes later Jesus was preaching on a hill farther along the fault-line. Just one minute has passed, out of Mother Earth's 46 'years', since man began his industrial revolution, three human lifetimes ago. During that minute he has multiplied his numbers and his skills prodigiously and ransacked the planet for metal and fuel.

A sketch of the Earth's history. The scale at the left shows the whole span of the Earth's existence from the time, thought to be 4600 million years ago, when the planet's surface ceased to be molten, following the formation of the Earth. The events occurring during this great expanse of time are being determined with increasing precision, including the origin of life and the principal mountain-building episodes. But the events of the past 600 million years, since the time when clearly recognisable forms of life became abundant, are much better known. A few of them are indicated on the right-hand scale, for the evolution both of the continents and of life. The scales follow the geological convention that puts the present at the top and the oldest rocks at the bottom – that is how they usually lie, under our feet. For the benefit of readers familiar with the names. of the geological periods, these are included for reference on the right-hand scale, but they are not generally used in this book.

Note. Abbreviations of the form −500 MY, −2 MY, etc. used throughout the text, mean 500 million years ago, 2 million years ago, etc.

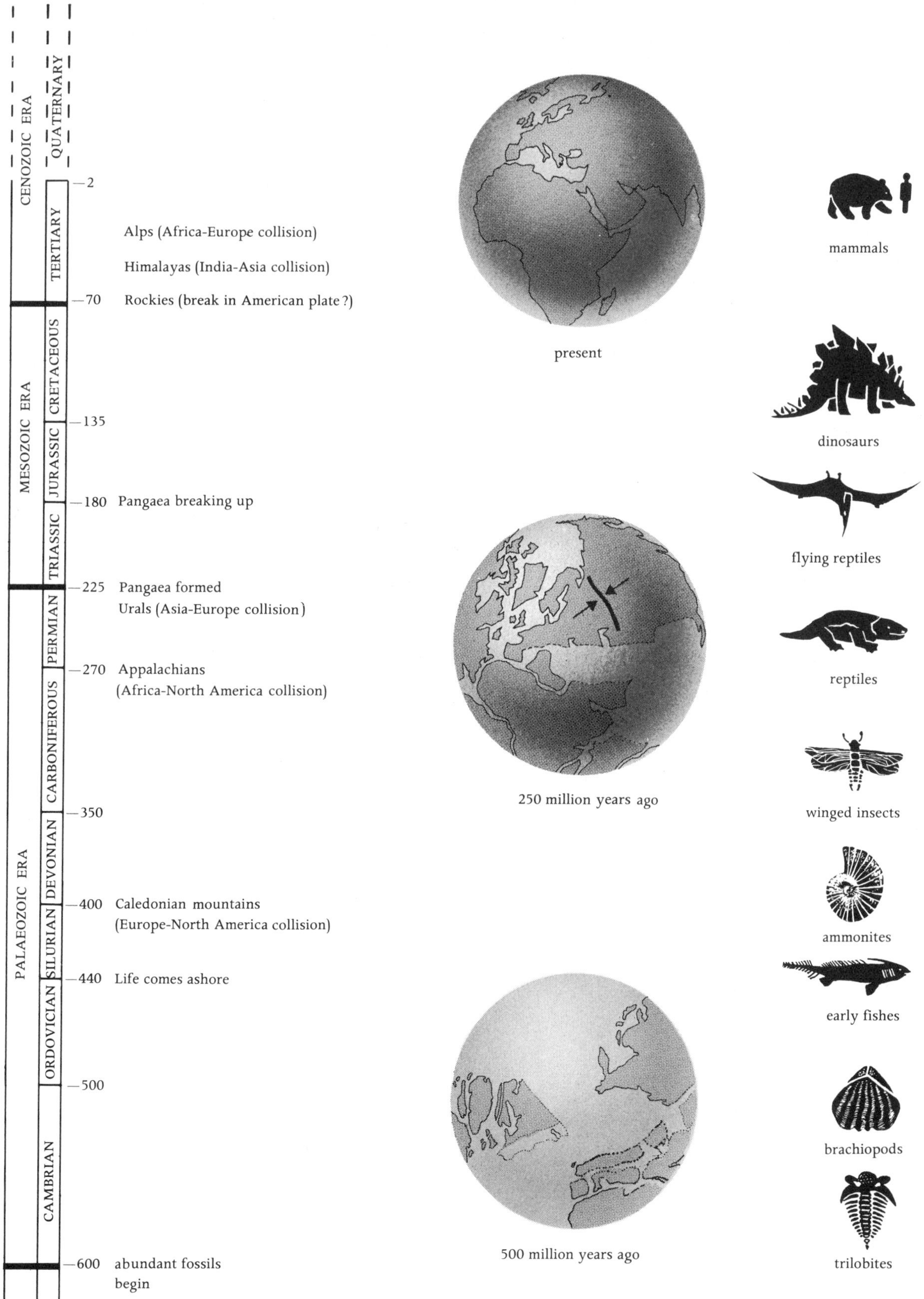

CENOZOIC ERA

QUATERNARY

—2

TERTIARY

Alps (Africa-Europe collision)

Himalayas (India-Asia collision)

—70 Rockies (break in American plate ?)

MESOZOIC ERA

CRETACEOUS

—135

JURASSIC

—180 Pangaea breaking up

TRIASSIC

—225 Pangaea formed
Urals (Asia-Europe collision)

PERMIAN

—270 Appalachians
(Africa-North America collision)

CARBONIFEROUS

—350

PALAEOZOIC ERA

DEVONIAN

—400 Caledonian mountains
(Europe-North America collision)

SILURIAN

—440 Life comes ashore

ORDOVICIAN

—500

CAMBRIAN

—600 abundant fossils
begin

present

250 million years ago

500 million years ago

mammals

dinosaurs

flying reptiles

reptiles

winged insects

ammonites

early fishes

brachiopods

trilobites

An underwater photograph (top) of a fresh fault in the floor of the Atlantic near Gibraltar. One mass of sediment has shifted upwards past another. This fault was found by French oceanographers soon after an earthquake in 1969. They also made soundings into the ocean bed (below) which reveal the full extent of the deformation of the Earth's surface, at the nearby Gorringe Bank. Two pieces of the Earth's outer shell are pressing together. They are making new underwater mountains, with a big depression to one side of them, revealed by the outline of hard rock (broken line). The figures 12h etc. record the time of day aboard the ship steaming over the bank; the figures 2s etc. show the travel-time in seconds of the sounding waves. The vertical scale is exaggerated but the new mountains are huge. (Centre Océanologique de Bretagne.)

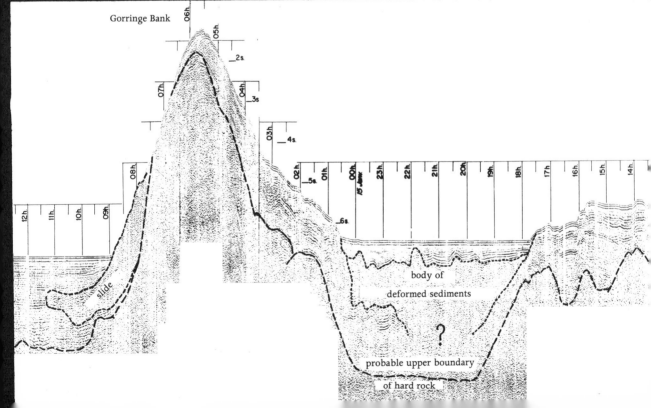

Gorringe Bank

slide

body of
deformed sediments

?

probable upper boundary
of hard rock

Chapter 1   A Disappearing Ocean

A new theory of the Earth accounts for earthquakes, volcanoes, mountain-building and the formation of minerals by one comprehensive process: the movement of huge plates of our planet's outer shell. In the Mediterranean region these events have occurred spectacularly, in the gradual destruction of an ocean which continues today.

The great Lisbon earthquake of 1755 shook Christendom morally almost as much as it shook the ground of Portugal, not least because devout Lisboners were in church at the time, celebrating All Saints' Day. The stonework fell on them; of the survivors, many fled to the waterside only to be drowned in great waves that raced in from the Atlantic. The event provoked impassioned preaching on the punishment of sinners and also anxious discussion of 'the problem of evil' by philosophers and theologians. Later generations adopted more scientific attitudes to earthquakes and fewer people were inclined to regard them as wrathful acts of God. Nevertheless, the mechanism of the Lisbon earthquake has become clear only in the past few years.

In February 1969, another earthquake caused damage and loss of life in Portugal, Spain and Morocco. Modern instruments pinpointed its origin near a range of submarine mountains, the Gorringe Bank, which runs out into the Atlantic from the mouth of the Mediterranean. French earth scientists from Brest went to the scene in the research ship *Jean Charcot*. They lowered a camera to the ocean floor and the photographs revealed a new break in the floor of the ocean. The rocks on one side of the fault had moved upwards by four feet. 'Mountain-building before our very eyes' was how Xavier Le Pichon, the leader of the research team at Brest, described it to me.

Although the Gorringe region was almost certainly the source of the worst of the earthquakes that affected Lisbon, on that grim All Saints' Day of 1755, the discovery of a new fault did not itself illuminate the cause; after all, everyone knows that the ground breaks in earthquakes. Closer investigations of the Gorringe Bank by the French group were more significant. They fitted the theories of Le Pichon and the other young earth scientists who, in the late 1960s, began to re-

interpret the changes in the surface of our planet in terms of one simple, all-embracing idea.

The idea, called 'plate tectonics', is that the outer shell of the Earth is divided into a small number of 'plates' and major changes in the Earth's surface occur only at the boundaries between plates. The plates fit closely together, yet they are moved around by internal forces that extend the plates in some regions and destroy them in others. The outlines of the plates are deduced from the zones at which earthquakes most frequently occur. One of these plate boundaries runs through the Gorringe Bank.

Jean Charcot explored the underwater mountains with shock waves transmitted from the ship to the ocean bed. Apparatus like a high-powered echo sounder detected not only the bed of the ocean but also the layers of mud and hard rock underlying it. By this means, Le Pichon and his colleagues found that thick muddy sediments hid the base of the Gorringe Bank. Subtract those, and the bank is seen to tower above a trough that has buckled downwards. The total height from trough to crest is seven miles; in comparison, the world's highest mountain, Everest, stands somewhat less than six miles above sea level. For Le Pichon, this is evidence that the plate carrying Europe and the plate carrying Africa are pressing together and squeezing up new mountains – in the process, causing the earthquakes.

The plate boundary between Africa and Europe extends right along the Mediterranean and makes it an excellent region for a first reconnaissance among the new concepts of the earth sciences. Many events afflicting the cradle of western civilisation have, until now, seemed arbitrary and unconnected. They all turn out to be inevitable results of the closing of the Mediterranean as the continents on either hand converge. The earthquakes and volcanoes mark a systematic remodelling of the face of the Earth. Near the scene of the great Sicilian earthquake of 1908, which killed 30,000 people in Messina, Mount Etna was in eruption once again in 1971; this volcano is part of the same story. So is Thera, at the eastern end of the Mediterranean.

The legend of Atlantis came to the Athenians from the Egyptians, who told of the overnight disappearance of an island off Greece. The Athenians themselves were said to be descendants of the warlike people who had lived there. All that was left of Atlantis, according to the legend, was a handful of islets. Marine geologists delving into the seabed and archaeologists digging for ruins are now inclined to identify small islands, Thera and its neighbours, whose ashy cliffs rise almost barrenly from the Aegean Sea, as the remains of Atlantis. There, early in the fifteenth century BC, an important trading city stood on a large volcanic island.

The first disaster was a series of eruptions that buried this nameless city like a prior Pompeii. In the second disaster, the island blew up one summer's day with a violence far surpassing even the notorious explosion of Krakatao near Java, in 1883. Thera sent towering waves of water breaking over coastal settlements hundreds of miles away and it may have delivered millions of tons of wind-borne dust or poisonous vapours to the cities and fields of Minoan Crete. Some investigators even try to link Thera with the Biblical plagues ('. . . and there was a thick darkness in all the land of Egypt three days'). The explosion was by any reckoning a turning-point in prehistory. It caused a shift of people and power from Minoan Crete to the mainland of Greece.

We can now specify the cause of the Thera volcano, which is still active from time to time. Southern Greece, Crete and the Aegean islands are riding on a small plate, a chip of the Earth's outer shell, which is travelling south-westwards towards Africa. It can do so only by overriding part of the floor of the Mediterranean, a deep oceanic basin. That ocean floor, and the African plate of which it is a part, is bent downwards, creating a pronounced trench of deeper water off the outer coast of Crete. The ocean floor is driven at a steep angle into the body of the Earth, where it is gradually destroyed. But, in the process, the enormous friction causes earthquakes and also generates heat that melts rocks and throws up volcanoes at some distance behind the trench.

One of the Bronze Age cities overwhelmed by the great waves from the explosion of Thera was Elaphonisos, at the southern tip of the Greek mainland. The impressive ruins of this seaport were discovered in 1967, by Nicholas Flemming, a British scientific diver. It now lies under water; the coast of Greece in this region has evidently subsided by ten feet during the three millennia since the town was destroyed. This deformation of the surface of the land is, in Flemming's view, another aspect of the slow extinction of the Mediterranean.

Flemming, from Britain's National Institute of Oceanography, is an indefatigable explorer of drowned seaports. He continues diving to investigate them despite an accident that has left him paralysed in both legs. Each sunken port has its own historical interest, but Flemming is especially concerned to measure the extent of the submergence, usually by seeing how deep the wharves now lie, and to compare results from many ports.

At Misenum, across the Bay of Naples from Vesuvius, the jetties of a port built by the Romans lie under several feet of water. A covered passageway through which sea-captains of Augustus once walked now forms a submerged tunnel for divers; it leads to a courtyard where sea water fills an ancient fish-pond. Around the bay are sunken ruins of other seaports that

have disappeared from sight under the waves. One conceivable explanation, that the sea itself has risen to drown the ports, is ruled out because in many parts of the Mediterranean the sea level is much the same as it was in Roman times. Elsewhere around the same shores, the land at ancient seaports has actually risen. These changes tell of the slow distortion of the ground in the jaws of a vice between Europe and Africa. They are testimony from the short span of human history of a great process which is taking many millions of years to complete.

The present convulsions of the Mediterranean are simply the second act in a drama in which geography has run riot. Europe and Africa have tussled for territory across a lost ocean, Tethys. In the process, many portions of land have been torn loose and re-shuffled to make the intricate pattern of promontories and seas in the Mediterranean, and incidentally to provide man with important supplies of minerals. At the climax to Act One Italy, a former part of Africa, was driven violently into the coast of Switzerland, thereby piling, folding and compressing material of the sea floor

to make the Alps. In Act Three, if present motions continue, an even greater chain of new mountains will replace the Mediterranean.

No tales of Olympus or Valhalla are bolder than those that earth scientists relate today. They speak of fissures in the shell of our planet that grow to make mighty oceans. Elsewhere, one section of the ocean floor destroys another and conjures new land into existence. Subterranean agencies shift whole continents around like flotsam. The land survives, but it carries the scars of crushing and tearing in repeated accidents which culminate, every so often, in a collision of continents. Such events would be hard to credit if one piece of evidence after another were not falling into place.

The horror of earthquakes and volcanoes has not been dispelled, but there is no longer uncertainty about their causes. In every part of the world, earth-quakes and volcanoes can now be explained by the interactions of the plates of the Earth's outer shell; so, too, can the creation of mountains and minerals. For the citizen of Tokyo or San Francisco anxious about the earthquakes threatening his city; for the Chilean or

The principal regions where earthquakes occur nowadays are shown on the uppermost globes. They form, for the most part, a web of narrow bands or lines across the face of the globe; between them are huge areas where earthquakes are rare. This pattern of earthquakes is the key to the present structure of the Earth's outer shell, as shown in the next map.

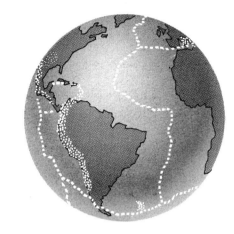

The master-map of the earth sciences today, shown on the lower row of globes and on another projection below. The map reveals the 'plates' into which the shell of the planet is at present broken. They are moving in different directions, carrying the continents on their backs. The plates are rigid; major geological events are mostly confined to the boundaries between the plates. The region labelled Ir (Iran) seems to be a 'soft' region with earthquakes throughout it, and earthquakes are also widespread in China.

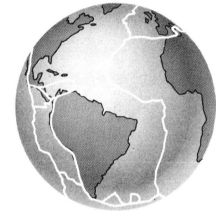

Major plates:
Af African
Am American
An Antarctic
Eu Eurasian
In Indian
Pa Pacific

Minor plates include:
Ae Aegean
Ar Arabian
Ca Caribbean
Na Nasca
Ph Philippine
Tu Turkish

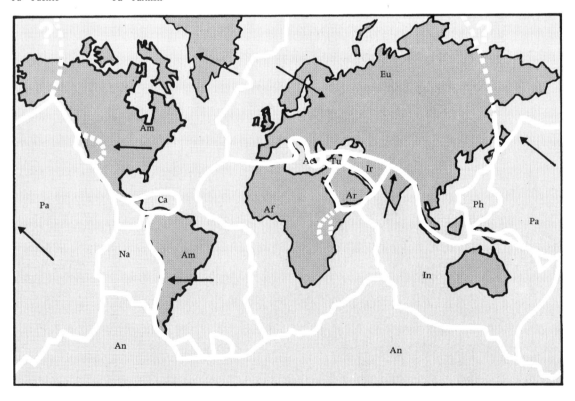

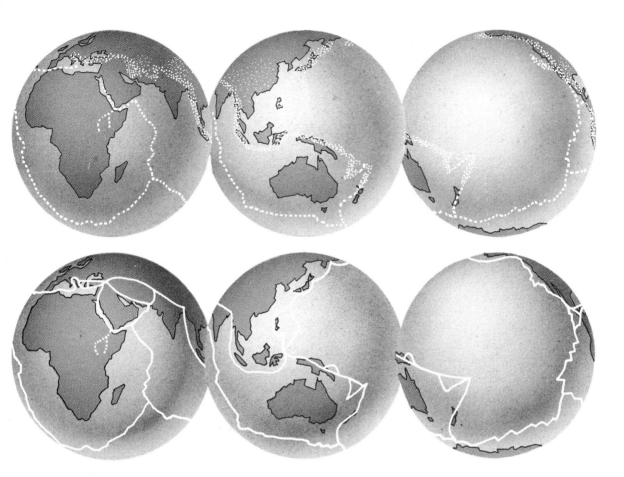

Icelandic farmer working on the slopes of a rumbling volcano; for the inhabitant of one of the quieter parts of the world who is simply curious to know why the ground he stands on exists or how there comes to be oil in the Arctic – for all who inhabit this planet, the earth sciences now supply a new enlightenment, tantamount to a rediscovery of the Earth. For the first time we can look on our familiar hills and tell why they came into being.

Facts in search of theory

'Above the plains of Italy where flocks of birds are flying today fishes were once moving in large shoals.' Leonardo da Vinci, in the fifteenth century, glimpsed the significance of the fossil seashells to be found on mountainsides. Long afterwards, slower-witted men were still contending variously that the fossils were no more than evidence of Noah's Flood or that pilgrims or the Devil himself had carried the seashells up the mountains. Like evolutionary biology, the sciences of the Earth took a long time to escape the circumscription of the Book of *Genesis*.

In the late eighteenth and early nineteenth centuries, sound geology emerged out of controversy. Men gradually came to accept that some of the hills over which they walked were actually extinct volcanoes although, in Freiburg, Abraham Werner asserted that the volcanic rocks had crystallised from sea water. In 1785, James Hutton of Edinburgh made explicit the cardinal idea that the face of the Earth was shaped over extremely long periods by processes of accumulation, heating, folding and erosion, still going on. Another Scot, Charles Lyell, born in the year of Hutton's death, had to say it all over again. William Smith, a canal engineer, used fossils as guides in mapping the strata of Britain, and specified that overlying rocks are normally younger than underlying rocks. In Paris, Georges Cuvier was finding systematic changes in fossil animals from one layer of rock to the next above. In 1840, Louis Agassiz of Switzerland discovered the evidence of past ice ages that buried much of Europe. Succeeding generations of geologists made steady and systematic progress, accumulating vast knowledge from many parts of the world.

Many dedicated and accomplished men seemed all

too ready to spend their lives scanning small parts of the Earth's surface in painstaking detail, but with little sense of global processes that govern events at any one place.

The scope and limitations of geology as it existed in the 1950s, before the present revolution, can be summed up as facts in search of theory. The old geology had become curiously dispirited, despite a great advance in the twentieth century, when atomic methods of dating rocks and geological events became available. There was mastery over 'what happened' but nothing but mystery about 'how it happened'. Geologists dwelt lovingly on the sculpture of landscapes by water, ice and wind. They could distinguish with facility between sediments of different ages incorporated in the rocks and tell whether they had been laid down on the land or under fresh or salt water, and in what climatic conditions; yet coal in Antarctica and evidence of ice sheets in India made no sense at all. The ocean basins were known to be very deep and to constitute huge depressions in the Earth, floored with heavy rock; but as to why they differed so much from the continents, there were only guesses. The locations of volcanoes seemed almost haphazard, although patterns were clearly discernible, including the 'ring of fire' surrounding the Pacific.

Especially puzzling was the existence of mountain chains like the Himalayas, the Alps, the Appalachians and many older 'fold belts' in all continents. Here, classical geology could certainly tell 'what happened', in the sense of describing a sequence of events. In several mountain chains, rocks that were formed underwater have been pushed high into the air. These sedimentary rocks accumulated great thicknesses before they were uplifted, suggesting that the floor of a shallow sea was sagging steadily lower to make room for the sediments. Why it sagged, who could say? Then

new forces of unspecified origin came into play and pushed the mountains up, often corrugating them to the point where the folds toppled sideways.

Somewhere into this story there fitted the idea that light, continental rocks eventually found their own level as they floated on the heavier rocks of the Earth's interior. Other notions were offered in various explanations of the sequence of events; they included currents in the Earth's interior and even a wholesale shrinkage of the planet so that the surface became like a dry apple's.

The puzzle remained, except among the few far-sighted geologists who knew that the continents had moved. They saw that colliding land masses would trap marine sediments between them and squeeze them to make new mountains. But the proof was lacking for such improbable motions of the solid Earth. Only in the mid-1960s did the evidence for the movements of continents and for the creation and annihilation of ocean basins become almost incontrovertible. Even in the 1970s, a handful of leading earth scientists remain firm in their denial that the continents have moved. Nor was it until 1967–8 that the new theory of plate movements appeared in precise formulation. That is recent indeed for a field of human inquiry that spans the whole extent of our planet, in time and space. The task of retelling the history of the Earth has only just begun.

This book is therefore an account of rediscovery in progress. It shows a reinterpretation of the old knowledge acquired by previous generations of geologists; only now can the real value of all that labour be seen. Suddenly, geology makes sense and, I trust, becomes more easily communicable to outsiders. The following chapters also tell of new knowledge acquired by powerful techniques – by drills that grope through three miles of sea water to sample the ocean bed, by detectors that catch the ghostly record of past magnetic events on the planet, and by earthquake 'telescopes' that look into the great body of rock far under our feet and find a graveyard of vanished ocean beds. More remarkable still is the way the ideas of an inspired young generation of earth scientists are turning out to match the way our planet actually behaves.

A broadside of continents

Two hundred million years ago, when the dinosaurs were in their ascendancy over the other animals, all the major continents of the Earth were locked together like pieces of a jigsaw puzzle. They made up the super-continent of Pangaea, the shape of which can be drawn with some confidence. Computers help in finding the best geographical fits between the present continental fragments. The results so obtained are checked by other means: for example, by evidence of climatic changes, by the past motions of the fragments implied by magnetism preserved in the rocks on land, and by deductions about the growth of present-day oceans.

The map of the supercontinent shows a great 'V', converging at the joint of southern Europe and north-west Africa. The ocean within the 'V' is Tethys. Asia makes up one arm of the 'V'; Antarctica and Australia extend the other, together with India, far removed from the rest of Asia. Nailed to the outside are the Americas, fitting the present coasts of Europe and west Africa. The motions, and the creation of new oceans, necessary for shifting the continents to their modern positions, are fairly obvious, but investiga-tions are now giving a timetable for the stages of the break-up and plotting the courses of the pieces in their long journeys across the face of the globe. The conti-nents are not self-propelled; they are carried on plates

about forty miles thick, which usually include portions of the floors of oceans. Indeed, as a continent moves, new ocean floor has to form in its wake, by the growth of the plate on which it rides. The plates, in turn, are driven by slow-flowing currents of hot rock deep inside the Earth.

The two arms of the 'V' of Pangaea have their own names: the southern arm is often called Gondwanaland, while the northern continents of North America, Europe and Asia together make up Laurasia. The break-up of Pangaea required the formation of great rifts through the supercontinent and these were accom-panied by volcanic eruptions that help to date some of the events. The rupture evidently began in the Gulf of Mexico and between North America and Africa, around —190 MY.

Thereafter, the continents unstitched themselves one by one from Pangaea. South America parted from one side of Africa, Antarctica from the other. There is most confidence about the later phases of the move-ments, because the corresponding growth of the Indian and Atlantic oceans is known in some detail. The Atlantic opened hesitantly; indeed, one of the last big events in the break-up was the rifting of North America from Europe. Another, far away, was the separation of Australia from Antarctica.

Because the continents are confined to the surface of a sphere, they cannot move apart without tending to suffer new encounters. North and South America were reunited quite gently, by the Gulf of Mexico opening like a folding door. Much more remarkable was the end of India's long journey across the Tethys Ocean. It helped to destroy the old floor of Tethys and opened the Indian Ocean behind it. Then India came into violent collision with Asia at about —50 MY. The results are plain to see, in the mountains made by that encounter – the world's greatest, the Himalayas. The

The supercontinent of Pangaea. The continents fitted together along the edges of the present continental shelves, so there were no gaps in the Atlantic region. The continental pieces of Pangaea have survived, although fragmented; the ocean floor, on the other hand, has almost completely renewed itself. This is A. G. Smith's fit of the continents, slightly modified.

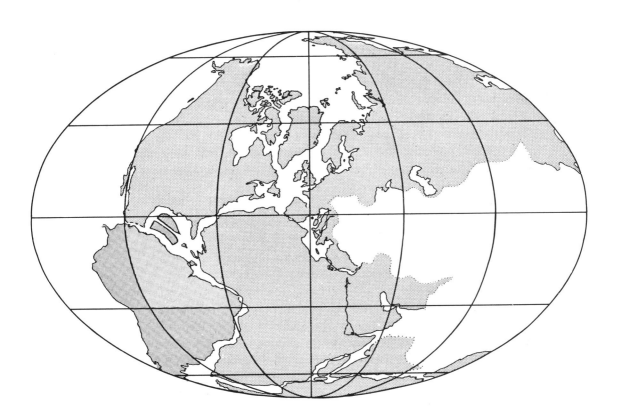

collision is still in progress, because India has not quite come to rest.

Everest itself is only the chief among giants; more than a hundred peaks in the Himalayas and neighbouring ranges stand higher than anything that the Andes can offer. Beyond the main line of peaks lies Tibet, a lonely platform a thousand miles across and as high as Mont Blanc. How the land ever came to be heaped so high is inexplicable except by the doctrine of moving continents. The great mountains are the product of crushing, buckling and heaping of material scooped from the ocean bed and the edges of the Asian and Indian continents, all along the colliding coasts.

The new narrative clears up a good deal of mystery about the Himalayas. For example, north of the high range, rocks are rich in the shells of ammonites, sea animals which flourished 150 million years ago, while the corresponding rocks to the south are almost bare of the corresponding fossils. To explain this difference, classical geology had to imagine a pre-existing barrier called the Himalayan ridge, much older than the present mountains, which prevented the ammonites from travelling south. Now the explanation is more convincing: the southern rocks of the same age were formed far away across Tethys.

Pieces of Gondwanaland have struck the hulk of

The disintegration of Pangaea. The broken lines show new or incipient rifts. Note the narrowing of the Mediterranean and the quick movement of India from Africa to Asia.

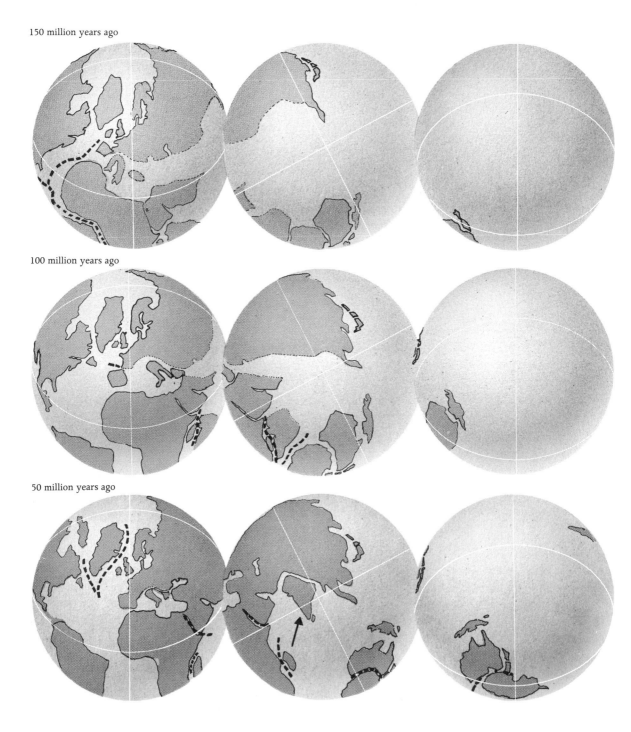

150 million years ago

100 million years ago

50 million years ago

The Earth today, showing how a new combination of continents is beginning to form, marked by mountains built in the collisions of Africa, Arabia and India with Eurasia.

Eurasia like a broadside: in the east, India; in the centre, Arabia and Iran (which broke from Africa at about —20 MY); in the west, Africa itself. This 'fall of shot' has created an almost unbroken series of new mountain chains running from Spain to China. Here and there, along this Alpine-Himalayan belt, are found peculiar collections of rocks, scraps of material pinched from deep inside the ocean floor, which are visible remains of Tethys. There are other reminders. Under this line of mountains, oceanic plates that have been swallowed in the collisions continue to sink into the body of the Earth, causing deep-seated earthquakes.

A surprising feature of the Himalayas is the way rivers such as the Arun, which runs between Everest and Kanchenjunga, have managed to saw right through the huge mountain range, from Tibet to the Ganges. These rivers must have existed before the mountains, and they show how gradually the mountains formed. As the Asian and Indian continents pressed together, these rivers cut the rock quickly enough to hold their own, despite the mountains rising in their path.

The motions of plates and of the continents they carry on their backs figure repeatedly throughout this book. These motions are slow by human standards, being typically in the range of a half to four inches a year. Greece, which is on one of the faster-moving plates, has travelled 100 or 200 yards since the time of Socrates. But time is one resource of which the Earth has plenty and the cumulative effect of these slow motions can remodel the Earth in a mere 150 million years, create great mountains, or knock down man's cities.

When one hears of collisions between continents, the inevitable mental picture of colliding cars is not entirely inappropriate. Mountains can be every bit as crumpled and confused as the wreckage of a bad road accident. Visualise it, though, as a crash in a slow-motion movie. When the continents make contact they do so more slowly and gently than a creeper grows on a wall, but they do not stop. They go on moving together inexorably. A remarkable feature of the process is the way it is driven to completion. Even though the colliding continental margins are irregular and poorly matched, they squeeze the intervening ocean completely out of existence.

After millions of years the forces that shuffle the plates around are finally beaten by the growing resistance of heaped-up rock, as the continents weld themselves together. Only then do the plate boundaries and motions change and the Earth finds an easier outlet for its energy. South of Ceylon the Indian plate is beginning to break; unable to push India further, the ocean floor must now dive out of the way.

Italy strikes home

As the Atlantic gradually opened, Europe remained tied to North America for longer than Africa did; as a result Africa and Europe did not move in concert. Their relative motions deduced from the growth of the Atlantic refer to the west coasts of those continents. The main action, though, occurred between the facing coasts of the two continents, in the lands bordering first the Tethys Ocean and now the Mediterranean. Spain swung around during part of the Atlantic development, but the events involving other Mediterranean territories and oceanic fragments have to be inferred by different kinds of evidence.

For example, geological information read in the modern way tells that the Alps were formed in collision between Italy and the rest of Europe; similar reasoning can be applied to other Mediterranean mountain belts. Deep-sea drilling gives facts about the history of the ocean basins within the Mediterranean. So do rocks on

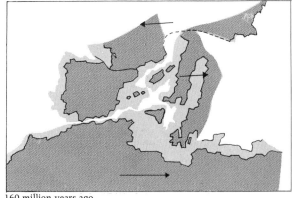

Relative movements of Africa and Europe that shaped the Mediterranean and made the Alps. (After K. J. Hsü.) The photograph below shows some of the wreckage of the collision between Italy and Europe.

160 million years ago

land that were magnetised by the Earth's field at the time of their formation; they show that the island of Sardinia, for example, has rotated through about 50 degrees during the past 200 million years.

There is also the jigsaw-puzzle approach, that sets out to fit the pieces of continents together in a reasonable-looking way. The first attempts at combining all these sources of information to give a narrative about the Alps and the Mediterranean were made in 1971 by Kenneth Hsü and by Alan Gilbert Smith. Hsü is a Chinese-born, American-trained geologist now working at the technological university in Zürich; Smith, at Cambridge University, has played a prominent part in reconstructing the supercontinent of Pangaea. Although they worked independently, they deduced the same sequence of events.

Before the opening of the Atlantic began, Africa and Europe were firmly joined in the west by quite a broad bridge. It contained, massed together, the land that was to become Greece, Yugoslavia, Italy and the islands of Corsica and Sardinia. At that time France was entirely landlocked, sealed in by the African bridge to the south, by Spain to the south-west and by Britain and Greenland to the north-west.

Smith also fits Turkey against North Africa at the start of the break-up, but he qualifies this by saying that Turkey is complex territory with slices of continental material sandwiching the remains of old ocean floor. He thinks that for an impression of the origin of Turkey one should look at the map of south-east Asia and Indonesia, with its jumble of islands and promontories, and then imagine them heaped together by an advancing continent. Persia, too, may have the same sort of history.

When the sidestepping dance of Africa and Europe began, Africa swept eastwards relative to Europe. The bridge splintered and part of it – Greece, Yugoslavia

and Italy – travelled with Africa. Eventually this great nose on the north of Africa struck eastern Europe, raising the Carpathian mountains that zigzag through the Balkans. The same process wrenched the Graeco-Yugo-Italian 'nose' and swivelled it anti-clockwise. Then, when the motion altered and Africa moved west and north in relation to Europe, it drove Italy like a ram into the middle of Europe. The opening of the North Atlantic between Europe and North America marked the change in the direction of motion.

If this reconstruction is correct, then all of the Mediterranean has been swept by the passage of these microcontinents, so that none of its oceanic floor can have been borrowed from the old Tethys Ocean. In that case, the only remaining candidate in the area that might be a possible scrap of Tethys itself is the Black Sea, between Turkey and Russia.

Turkey in the scissors

Dan McKenzie of Cambridge University is one of the leading theorists of plate tectonics. He was only twenty-five when, in 1967, he arrived at some of the guiding principles about the behaviour of the rigid plates that jostle together at the Earth's surface. The picture at that time was clear only for the oceanic regions of the Earth's crust, where the boundaries between plates were sharply defined. Since then McKenzie has been greatly exercised by the question of how useful the new ideas will be in describing what happens on land, where the patterns of earthquakes show that the Earth is changing over much wider zones and the plate boundaries are fuzzy. Despite their greater bulk, continents are frailer than the floors of the deep oceans and the old scars of past upheavals that they carry are liable to start slipping in many places in times of stress – as when continents collide.

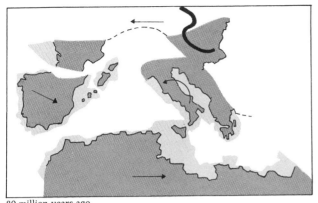

80 million years ago

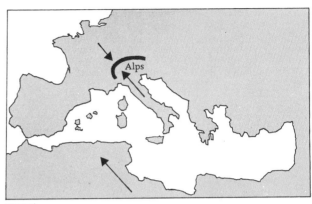

present

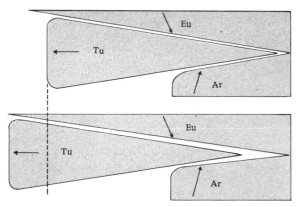

The region where Eurasia, Africa and Arabia are coming together, in the Mediterranean and the Middle East, seemed to McKenzie to be an excellent testing-ground. He set out to see whether the new ideas could make any sense of the apparent chaos in the region.

The chief method he used depended on interpreting the patterns of initial tremors from earthquakes. An earthquake occurs along a fault, a crack in the Earth's shell, where the rocks on the two sides are trying to move past one another. They suddenly succeed in doing so; one mass of rocks lurches in one direction, and the other in the opposite direction. The first pressure wave arriving at a distant seismic recorder will be either a push or a tug, depending on whether the ground on the near side of the fault is moving towards the station or away from it. By comparing the 'first motions' at a number of stations, the main direction of movement in the earthquake can be discovered. This technique was of exceptional importance in proving the general motions of plates around the Earth. If several earthquakes along a plate boundary show the same direction of motion, they provide evidence that the plates on either side are moving as rigid units.

This, then, was the technique adopted by McKenzie for a reconnaissance of the plate movements in the Mediterranean region. In 1970, he diagnosed the action of two small plates, in addition to the Eurasian, African and Arabian plates already known to be involved in the area. He called them the Aegean plate and the Turkish plate. The Aegean plate, carrying southern Greece and Turkey, has already been invoked to explain Thera. It is travelling southwestwards, tending to tear Greece apart and creating a rift behind it in the northern floor of the Aegean Sea. The Turkish plate is travelling west, pushing the Aegean plate. One of its boundaries is the Anatolian fault, which runs through northern Turkey and is a source of many terrible earthquakes. The rate

of motion of the southern part of Turkey along the Anatolian fault has been estimated at $4\frac{1}{2}$ inches a year.

What is the reason for this hasty westward journey of Turkey and Greece, while the motion between Eurasia and the main southern plates is roughly north-south? McKenzie concedes that the motion is 'complicated and surprising' but he thinks it has a simple explanation. Turkey is caught between the Eurasian and Arabian plates as if in a pair of scissors, the hinge of which is in the east. As the blades close, Turkey and Greece are free to move out of the way by overriding the oceanic floor of the Mediterranean. But they have to move quickly – for a slight closing of the scissors, the small plates must travel a comparatively long way if they are still to find room enough between the blades.

South from Turkey there runs a depression in the Earth's surface that is redolent with human history: the Jordan Valley, with the Dead Sea and Lake Tiberias (the Sea of Galilee). It has often been described as a rift valley – a slight tearing apart of the Earth's surface – but more careful investigations have shown it also to be a fault line along which Palestine has slipped 70 miles southwards in relation to Jordan during the past 20 million years. It seems that Palestine moved with Sinai, in making a minor adjustment to the motions between Africa and Arabia, associated with the birth of the Red Sea.

Plates and chronicles

Further east, in the Caucasus and Persia, the Arabian and Eurasian plates are pressing hard together, making mountains and thickening the Earth's crust. The earthquakes become widespread and complicated, because of the frailty of the continental material and the re-activation of ancient faults.

Nicholas Ambraseys of Imperial College, London, is

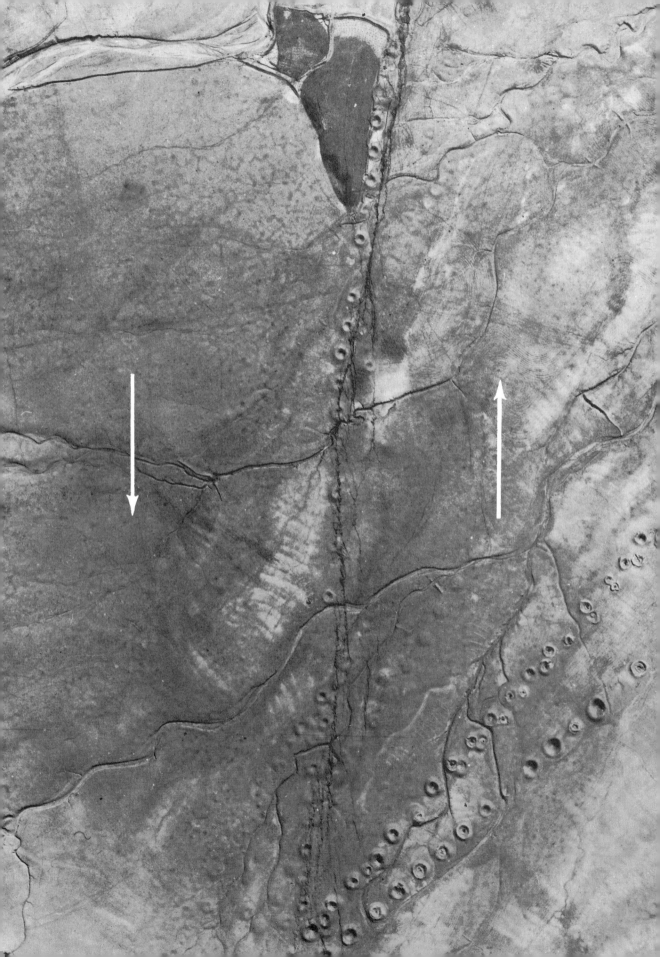

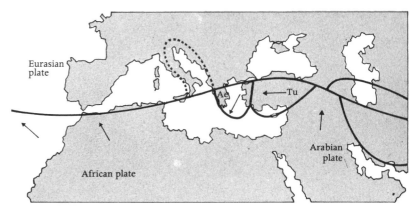

Plates of the Mediterranean region, showing how southern Greece (Aegean plate, Ae) and Turkey (Tu) are pushed sideways. The dotted line indicates a probable extension of the African plate, to northern Italy. (Adapted from a map by D. McKenzie.)

Eurasian plate

Ae

Tu

African plate

Arabian plate

an engineering seismologist whose concerns are mainly practical, in trying to help the inhabitants of countries like Persia to understand their earthquakes and to take precautions against them. The new fundamental studies are entirely relevant to these purposes.

Most of the studies are based on earthquakes occurring in the past few years. While the plate boundaries thereby defined help to show which areas are at risk and why, they may be misleading if some boundaries have recently been inactive. In other words, there may be quiescent plate boundaries that could start working again, even in areas at present thought to be earthquake-free. Ambraseys has therefore been leading an international investigation that combines geology with historical scholarship. The aim has been to map the occurrences of earthquakes in the eastern Mediterranean and the Near East, going back to the first century AD.

The work has been slow. It required the reading of chronicles and manuscripts in many ancient and modern languages. These have been supplemented by clues from a variety of sources, including archaeological evidence of damage and even coins that were issued as relief after destructive earthquakes. By 1971, Ambraseys and his scholars had identified 3000 earthquakes occurring in the region between 10 AD and 1699. Of these, 2200 were major shocks.

The resulting maps of historical earthquake activity are very revealing, when they are compared century by century, or with the records of the twentieth century. While they do not show any plate boundaries unnoticed in contemporary seismology, they amply confirm Ambraseys' suspicion that the recent records do not tell the whole truth about which areas are at serious risk from earthquakes.

Some regions where few earthquakes occur nowadays have an ominous history. They are the so-called

Earthquake damage (1968) at the Persian village of Bidokt. It escaped relatively lightly; in other villages nearly every house was destroyed and many of the inhabitants were killed. The traditional methods of construction, including the use of heavy, domed roofs, are not at all well suited to an earthquake-prone area like the Middle East.

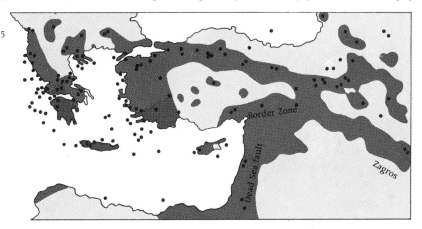

earthquake zones AD 10 to 1699
• major shallow earthquakes 1900–65

Border Zone (the south-east boundary of the Turkish plate), the Zagros mountain region of Persia, the Dead Sea fault through the Levant, and the shores of the Red Sea. It cannot be said that seismic days are over for these regions, because the historical patterns show that a zone can be quiescent for as much as 500 years and then begin to move in a new, long-lasting series of disastrous earthquakes.

Between 500 AD and 1000, there were many earthquakes in the Border Zone but the Anatolian fault across northern Turkey was relatively quiet. Then the Anatolian fault began moving again and is still very active today, while the Border Zone is quiescent. These changes say something about how the small Turkish plate moves, shifting first on one boundary and then on the other. That is a point of theoretical as well as practical interest.

For movements of human populations, too, Ambraseys' historical studies reveal a striking pattern. Villagers often decamped after an earthquake, building afresh on what they hoped would be a safer site; but the vested interests represented in cities invariably ensured the rebuilding of a devastated city at the very scene of its destruction.

'A hole in the Earth'

Glomar Challenger is a remarkable ship, equipped for drilling into the bed of the deep ocean and recovering evidence about the Earth's history. Since the end of 1968 she has operated almost non-stop, with alternate crews, in the Deep Sea Drilling Project of the US National Science Foundation. She does not seek oil or military advantage or any of the other prizes to which advanced marine technology tends to be dedicated; her quest is for fundamental knowledge. *Challenger* is for the ocean scientists what the microscope was for the early biologists, supplying information and edification obtainable in no other way.

In 1970, *Challenger* passed through the Gibraltar Strait. In a two-month cruise, she made twenty-seven borings at fourteen carefully chosen sites along the length of the Mediterranean. The full analysis of the ocean-floor material brought up on the end of the long drill string is not done overnight. No doubt geologists in the twenty-first century will still be examining the 'cores' from *Challenger*'s cruises, now kept in cold store at the Scripps Institution of Oceanography near San Diego and the Lamont-Doherty Geological Observatory near New York. Nevertheless, the chief scientists

A gorge cut through the Himalayas. The Indus river that accomplished this feat is older than the mountains. It sawed the rock quickly enough to hold its own when India collided with Asia and the mountains rose – at a few inches per century.

of that Mediterranean expedition in *Challenger*, William Ryan of the Lamont Observatory and Kenneth Hsü from Zürich, have been willing to talk of their preliminary impressions from the cruise.

Past events in the Mediterranean are revealed in the material of its bed. Vigorous currents from the Atlantic and Indian Oceans were sweeping through the Mediterranean 15 million years ago. By about —10 MY the region was cut off from both the other oceans, as the Mediterranean narrowed. Hsü and Ryan think the sealing-off was complete and that the sea water evaporated, leaving a thick coating of salts all over the ocean bed. Ryan has an extraordinary picture of a dried-up ocean – a 'hole in the Earth' three miles deep – and then of a great deluge when the western end began to reopen and the Atlantic water poured into the hole. Just inside the Mediterranean, between Spain and Morocco, *Challenger*'s corer brought up evidence that fierce erosion of the ocean floor occurred at that time.

Results of the drilling support the theory that the Mediterranean is even now being squeezed out of existence between Africa and Europe. For example, south of Greece and only about 150 miles from Thera-Atlantis, in a trench of deep water *Challenger* found young rocks driven underneath much older rocks – clear evidence that one piece of the ocean floor is being swallowed underneath another. Another discovery was that an important hump in the ocean bed, running from east to west along the eastern Mediterranean, carries mud from the Nile which is less than two million years old. The hump must therefore have risen very recently indeed, by geological standards, to climb out of reach of new sediments from the Nile.

At one end of this hump stands the island of Cyprus, just off the coast of Turkey. There, the deformation of the ocean floor has been so great that part of the land of Cyprus is a piece of it.

Taxi to the moho

Copper metal, which was a key ingredient of prehistoric bronze as well as of modern electrical machinery, takes its name from Cyprus. This, the most easterly island of the Mediterranean, is riddled with tunnels dug by Romans in search of the ore of *cyprium aes*; before the Romans, Cyprus had furnished the armouries of the Phoenicians with copper. But the island was stripped and today the slag heaps left by the Phoenicians are richer in copper than most of the remaining ore.

The copper ore of Cyprus formed at the bottom of the sea. Drilling by *Glomar Challenger* around the world has shown metal-rich sediments, including copper compounds, to be a common feature of the ocean floor. In some places the ocean floor even has a layer of copper metal built into it. When the southern part of Cyprus rose from the sea about two million years ago, it brought up copper ore with it.

The mechanism that raised Cyprus has seized the attention of the world's geologists. It was in 1963 that Ian Gass, then at Leeds University, and David Masson-Smith of the Overseas Geological Surveys reasoned that Mount Olympus, in the west of the island, consisted of rocks quite different from ordinary continental or ocean-island material – some being rocks that normally belonged much lower down, under three miles of ocean water and a further four miles of ocean-floor rock.

At around that time, the US Government was supporting the Mohole, a most ambitious project to drill for several miles into the bed of the ocean. This would have been far deeper than *Glomar Challenger* is now drilling and the aim was to recover samples of rock from the deep interior of the Earth. It took its name from the 'moho', a level in the Earth at which shock waves from earthquakes suddenly begin travelling faster, as if they have entered heavier rock of a different composition from the uppermost layers of the Earth.

The Matterhorn (left), a piece of Italy that has overridden Switzerland in a collision between Africa and Europe. It is a remnant of a much greater mass of rock largely eroded away in the 20 million years or so since the collision occurred.

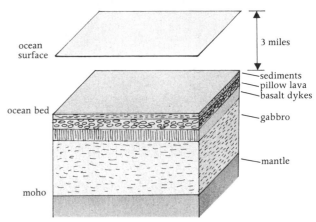

Ocean-floor rocks (right). Beneath the muddy sediments is hard rock, made successively of pillow-shaped lumps of lava, closely packed vertical 'dykes' of basalt, and the gabbro, a shapeless mass of rock. A change in speed of earthquake waves marks the moho, the transition from 'crust' to 'mantle'. All of these rocks are found on land in Cyprus, and some of them occur at the base of the Matterhorn.

The Mohole started as a scheme expected to cost five million dollars, but grew in difficulty until by 1966, when the Congress refused to finance it further, the expected cost was more than 100 million dollars. Many earth scientists were bitterly disappointed; others adopted the slogan for what became the *Glomar Challenger* programme: 'Not Mohole but more holes'. Maybe it is just as well that so much money was not spent on the Mohole, because anyone can recover samples from the deep interior of the Earth for the price of an airline ticket to Cyprus and a taxi to Mount Olympus. Perhaps they are a bit weathered, but not by 100 million dollars' worth of weathering.

Closer study of the rocks of the region reveals all the layers of the ocean floor through which the Mohole would have bored. The copper ore lies on top of the hard rock of the floor. Pillow-like lumps of lava, abundant in Cyprus, correspond to the uppermost layer of that hard rock. Then there are sheets of basalt packed as closely as playing cards in a pack; they are from the next layer down; and so on, until you come to heavy rock from beneath the moho itself.

In 1971, two expeditions of British earth scientists went to Cyprus to try to confirm that this identification of rocks was correct. The natural test was to see how fast shock waves travelled through the land of southern Cyprus. If much of the rock really came from beneath the ocean floor, shock waves from small explosions, picked up at distant detectors, should travel a good deal faster than they would through ordinary rocks. The first expedition turned the matter into a scientific cliff-hanger by reporting that the 'seismic velocity' seemed to be quite low. But the experiment was apparently done on too small a scale.

The second expedition included Fred Vine from the University of East Anglia, a man whose earlier investigations of the ocean floor had provided the crucial proof of the drifting of the continents (see p. 41). This time the charges went deep into boreholes left by mining companies looking for copper; the detectors were spaced up to 15 miles away. The distance was still too short for the experimenters to 'see' the densest rocks from beneath the moho. Nevertheless, the seismic velocities matched well with those to be expected in successive layers of the ocean floor.

That is not the end of the story, Cyprus is more than a geological curiosity. It illustrates a process that has occurred in several of the main mountain chains of the world, namely the incorporation into the land of these rocks from beneath the bed of the ocean. Where they show up, among mountains that often are far from the sea, they mark the former plate boundary at which the collision occurred that made the mountains. Colleagues of Vine's at East Anglia, J. A. Pierce and J. R. Cann, have found a chemical 'signature' in certain of these peculiar rocks which identifies them with the lavas of the ocean floor, even though their appearance may have changed drastically. The relative amounts of three chemical elements (titanium, zirconium and yttrium) varies in different kinds of volcanic rocks but they are the same in ocean-floor lavas, in material from Cyprus and in inland rocks in the Alps.

Africa for alpinists

The peak of the Matterhorn, rising to nearly 15,000 feet at the frontier of Switzerland and Italy, is not the highest in the region, but it has always been a special challenge to mountaineers in the Alps, because of its sharp profile. This pointed pyramid of rock was carved, by the action of ice, out of a much larger sheet of rock that formerly covered the area. We can now suppose, with considerable confidence, that the top section of the Matterhorn is a piece of Africa that was pushed on top of Europe. By

Africa we mean Italy, originally a part of Africa, which was torn, rotated and driven towards Switzerland.

Sandwiched between Europe and Africa, in the lower slopes of the Matterhorn, is material from beneath the floor of the ocean. This middle layer is much mangled and modified, but includes heavy rock of essentially the same kind as that found in Cyprus, pushed up from the floor of the Mediterranean. Similar rock occurring elsewhere along the Alps traces the line of collision between the continents.

If more attention had been paid to distinguished investigators of the Alps, the fact that the continents have moved might have been established long since. It was not with benefit of our present hindsight that they looked at the Alps and saw that they were made by some enormous horizontal pressure which ruckled the Earth's surface like a carpet. Alfred Wegener, the German meteorologist who offered the first comprehensive arguments for continental drift more than half a century ago, counted several Alpine geologists as his witnesses.

As long ago as 1829, Elie de Beaumont recognised the need for such a force to explain the Alps and he imagined that the Earth had shrunk, and its surface wrinkled. That theory became very popular during the nineteenth century. In 1922, Emile Argand seized on the proposition of drifting continents in explaining the Himalayas and the Alps and he even anticipated recent ideas by suggesting that Italy had swivelled before driving into Switzerland. Such was the power exerted over some geologists' imaginations by the overthrown rocks that constitute the Alps.

One present-day investigator, John Ramsay of Imperial College, London, is an expert on how rocks become folded. He finds special fascination in the way the distortions of whole Alpine ranges are matched by deformations on a much smaller scale within the individual bodies of rock of which the mountains are composed. In other words, one rock face can show a model of the whole range, on a scale of feet instead of miles. The rocks have been kneaded through and through, like modelling clay, rather than simply rearranged like bricks. Even under a microscope the rocks reveal the same patterns of buckling and looping that the mountains do. With layers of actual modelling clay, subjected to a very slow sideways pressure, Ramsay can reproduce the Alpine patterns in miniature, in the laboratory.

In real rocks, the reworking produces strange effects. A round pebble can be squeezed into the shape of a narrow stick; fossil animals adopt weird shapes as if in a distorting mirror. In a mountain range, the crumpling can reach the stage where folds are toppled forward, to lie recumbent on one side. Such kneading was not enough to complete the Alps; the strain eventually broke the rocks and, where that happened, broad sheets, or 'nappes', thrust their way forward for many miles over the top of other rocks.

These overthrusts often give themselves away to the geologist because they put old rock on top of younger rock – a most disorderly state of affairs, like finding a Neanderthal man buried on top of a Napoleonic soldier. One of the most conspicuous overthrusts is in the Glarus district of the Swiss Alps, where dark rocks about 250 million years old sit on light coloured rocks about 50 million years old. These ages have nothing to do with the date of the Alpine upheaval – they are simply the ages of some of the materials that were caught in it.

When the thrusting movement is retraced, it turns out to amount to many miles. If the Alps as a whole were straightened out, to rearrange the rocks as they were at the onset of the great collision, Europe would be at least 60 miles wider than it is; that is a measure of the squeezing of the Earth's surface.

The young Alps were heaped over the remains of

much older mountains, which the weather had worn almost flat since their formation at about —280 MY. Some of the old rocks were buried but others were squelched out to make new mountains, including Mont Blanc, which overtops the other peaks of the main Alps.

Those previous mountains, too, had run from east to west across Europe. They stretched from Poland, through Germany, France and southern England; the Harz mountains, the Ardennes, the jagged coasts of Brittany and Cornwall, all these are remnants of a former chain of Alps-like mountains.

But only half a chain. The other half lies on the far side of the Atlantic ocean in the younger Appalachians of the United States, broken off by the opening of the Atlantic. They were evidently formed in an even more formidable collision of continents: when North America was joined to Europe, and Africa ran into them both. The dinosaurs could have marched along that chain of mountains from Poland to Alabama, by way of Ireland and New York. Before plunging that far, and farther, back in time, we should arm ourselves better with the ideas and evidence of the new geology.

The Grand Canyon of the Colorado River. About 500 million years of the Earth's history is recorded in the successive layers of rock here exposed. During much of that time North America lay in the tropics, by the evidence of rock magnetism.

Chapter 2   Earth Rediscovered

Crucial evidence for the motions of continents and the growth of new oceans has come from weak magnetism frozen into rocks at the time of their formation. But the consequent theory of plate tectonics looks beyond 'continental drift' to a thorough explanation of geological changes now in progress around the world.

By a lake in south-eastern Australia about 26,000 years ago, people not unlike the present-day aborigines cremated the body of a young kinswoman. Today Lake Mungo is dry. In 1969, archaeologists from the Australian National University in Canberra identified the cremated bones as the oldest human remains so far discovered in the continent. Two years later, their earth-scientist colleagues made another remarkable discovery, in a fireplace just a few miles away on the same 'shore'.

In several countries, archaeologists and earth scientists have collaborated in studying the magnetism of old bricks. When a brick has been baked and it is cooling, it becomes faintly magnetised by the Earth's magnetic field prevailing at the time. Long afterwards, sensitive instruments can detect that fossil magnetism. If the bricks have not been moved, they preserve, 'frozen-in' as it were, the direction of the Earth's magnetic field at the time of their last firing. With help from the archaeologists, earth scientists can find out how the Earth's magnetic poles have wobbled around during thousands of years; in return, the archaeologists acquire another way of checking the consistency of their dates for human events.

A research student from Canberra collected samples from ancient hearthstones left by Australian aborigines around the countryside and brought them back to the laboratory. There was nothing unexpected about most of the hearthstone samples, many of which were less than 4000 years old. But some material from Lake Mungo, which was even older than the bones, gave a very odd result. It was magnetised in completely the wrong direction, as if the Earth's north and south magnetic poles had changed places.

For the leader of the research group, Michael McElhinny, an experienced rock magnetist, this was exciting news. 'Flips' of the Earth's magnetic field were well known, though most of them occurred millions of years ago. Like bricks and hearthstones, cooling volcanic rocks adopt the prevailing magnetism of the Earth and rocks of different ages tell of a long series of reversals. The last definite one occurred 700,000 years ago. But in 1967 French scientists had reported that the rocks of an extinct volcano indicated a brief reversal about 20,000 to 30,000 years ago. The aborigine hearthstones supply convincing evidence for it. The Australian results show that the magnetic poles swung into the opposite hemispheres 30,000 years ago. They probably stayed reversed for no more than 2000 years. That would not be surprising in view of a study of a reversal occurring 15 million years ago and recorded in fine detail during the progressive cooling of granite now exposed on the southern side of Mount Rainier in the state of Washington. Investigators from the University of Pittsburgh have found that, during the reversal, the magnetic poles vacillated between the two hemispheres for a thousand years, before settling in the 'flipped' position.

Nevertheless, the Lake Mungo fireplace confirms that a great geophysical event occurred during *Homo sapiens'* brief tenure of the Earth. It provides a fitting introduction to the story of how rock magnetism and the magnetic reversals helped to prove that the continents have moved and thereby opened the way to the present revolution in *Homo sapiens'* understanding of his planet.

Rock magnetism was for a long time the almost private interest of a few investigators in France and Japan. Volcanic rocks magnetised in the wrong direction, compared with the present magnetic field of the Earth, had been known since 1906 and for many years the French and Japanese researchers were pre-occupied with whether such rocks could spontaneously reverse their own magnetism. It was an important

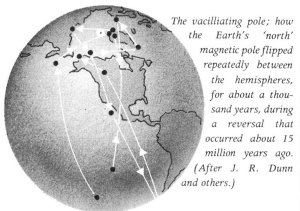

The vacilliating pole; how the Earth's 'north' magnetic pole flipped repeatedly between the hemispheres, for about a thousand years, during a reversal that occurred about 15 million years ago. (After J. R. Dunn and others.)

Excavation of an ancient fireplace at Lake Mungo, New South Wales. Material from this spot revealed an apparent reversal of the Earth's magnetic field about 30,000 years ago. In the background is a sample encased for removal to the laboratory.

question, to which the answer was 'yes, sometimes', but in the 1950s scientists in Britain began to interest themselves in rock magnetism and they scented bigger game than that.

They were aided by very sensitive magnetic instruments, including one designed by a famous atomic physicist, Patrick Blackett, to test an abortive theory of the origin of the Earth's magnetism. A much wider range of rocks was open to investigation. Keith Runcorn, a one-time radar scientist, led another group of rock magnetists, first at Cambridge and later at Newcastle-upon-Tyne.

Blackett's group had moved from Manchester to London when, in 1954, John Clegg, Mary Almond and Peter Stubbs announced that during the past 200 million years Britain had rotated clockwise through about thirty degrees and had also travelled a great distance to the north. They drew this conclusion from consistent misalignments in the magnetisation of sandstones dating from around −200 MY. Rocks that are formed cold, by the settling of grains, sandstone for example, are very weakly magnetised; there is simply a tendency for the grains to settle with their own slight magnetism aligned with the Earth's. Field reversals were present in about half the sandstone samples, but were not the point of interest. There were more subtle discrepancies in the directions of magnetism. The rocks were fossil compasses that told of the voyages of continents.

At about the same time, Ted Irving, formerly with Runcorn's group, was working in Australia and obtaining similar but even more striking results for the motion of that continent. To begin with, though, there was prolonged dissension in the ranks of the rock magnetists themselves. From Newcastle, Runcorn was arguing that the whole Earth could roll about, in relation to its axis of rotation, and that what the London group's results showed was not continental drift but

The Earth's magnetic field (a) endows newly forming rock with magnetism oriented northwards and dipping into the ground at an angle that increases with the distance from the Equator. The magnetism of old rock (b) tells of changes in orientation and latitude since its formation.

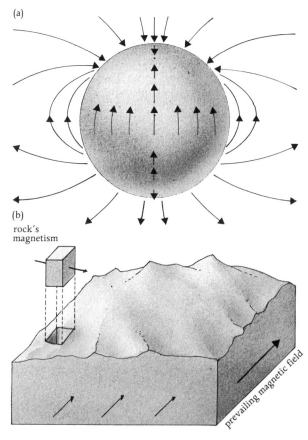

(a)

(b)
rock's magnetism

prevailing magnetic field

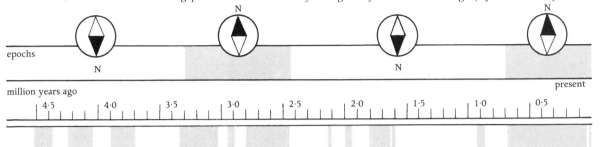

Reversals of the Earth's magnetic field during the past few million years. The field was 'normal', as today, during periods shown tinted; it was 'reversed' during periods shown white.

Note the broad epochs above and, underneath, the briefer reversals within them. The latter include the event of 30,000 years ago, confirmed at Lake Mungo. (After A. G. Cox.)

'polar wandering'. Only when his own group found different 'wandering' tracks for the poles, as deduced from rocks in different continents, was Runcorn convinced that continental drift had occurred.

By 1960, Blackett, Clegg and Stubbs were able to review the magnetic data from rocks and assert with confidence that since —500 MY the continents had moved north or south through great distances, at different rates, and had rotated through considerable angles. Americans for the most part remained unpersuaded; indeed although the Carnegie Institution in Washington had done important early work on rock magnetism, the subject was curiously neglected in the United States.

The riddle of the stripes

The rediscovery of the Earth has been a great collaborative enterprise of many experts in many branches of knowledge and from many countries. But, as is not unusual in science, one crucial idea transformed the whole investigation. It, too, involved the magnetism of rocks, but it has to be seen against a background of growing knowledge of the ocean floor.

Pioneer voyages during the nineteenth century, with deep-sounding cables, showed that the oceans were typically about three miles deep, but they also encountered the huge structure of the Mid-Atlantic Ridge and deep ocean trenches off Japan and elsewhere. When oceanography became better supported with ships and other resources, in the 1950s, new information accumulated rapidly. The Mid-Atlantic Ridge turned out to have a deep rift along its summit. Remarkably long, straight fractures were discovered in the floor of the Pacific. Perhaps the most provocative conclusion, from new charting of the ocean floor, was that a succession of great ridges, like the Mid-Atlantic

Ridge, ran through all the world's oceans, although not always in the middle.

Despite these discoveries there was, all through the 1950s, unease among the ocean scientists. The muddy layers on the ocean floor were very much thinner than you would expect if the oceans had been collecting sediments for most of the Earth's history. Harry Hess of Princeton University accepted the logical conclusion; he declared that the oceans were comparatively young. Many years before, Arthur Holmes had sketched a theory in which the mid-ocean ridges were places where hot material was coming up from the interior of the Earth while, at the ocean trenches, cold material was sinking back into the interior.

Around 1960, Hess elaborated and refined this theory, in the light of fresh knowledge, into a scheme of ocean-floor spreading. It said that new ocean floor was continually manufactured at the mid-ocean ridges, whence it travelled slowly outwards, eventually to be destroyed in ocean trenches.

During that period, ocean scientists were puzzling over yet another recent finding. Research ships zigzagging about the oceans had been towing very sensitive magnetic detectors which had revealed magnetic irregularities across the oceans. Over the Mid-Atlantic Ridge, for example, the Earth's magnetic field was about one per cent stronger than expected. In other zones, the field was slightly weaker than expected, and the changes from 'weaker' to 'stronger' or back again could occur in the space of a few miles. Maps of the north-east Pacific and the north-west Indian Ocean showed a curious stripy pattern. The ocean was marked by bands of alternately stronger and weaker magnetism. The pattern made no obvious sense at all. Ronald Mason, a British geophysicist working with the Scripps Institution in California, first discovered it, but could not explain it.

Aboard the drilling vessel Glomar Challenger, *looking forward over the rack of drill pipes. This ship is carrying out the scientific Deep Sea Drilling Project around the world's oceans. She can recover samples from beneath the ocean bed in water depths of up to 25,000 feet, being held in position automatically by the use of an acoustic beacon on the ocean floor.*

The tape recorder on the ocean floor (right). As rock moves away from the ridge crest, new rock coming up to fill the gap adopts the prevailing magnetism of the Earth. The floor on each side of the ridge records the erratic 'flips' of the Earth's field, in a series of stripes. Real magnetic stripes in the map to the right are from the Mid-Atlantic Ridge near Iceland. (After F. J. Vine.)

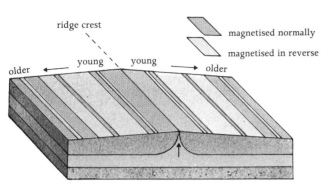

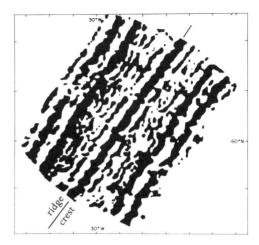

Fred Vine, a 23-year-old research student at Cambridge University, was familiar with these results. He also knew of Hess's theory of ocean-floor spreading. That theory appeared far less 'wild' in British eyes than to many of Hess's fellow Americans. Even so, there seemed to be no easy way of testing it until, one day, Fred Vine saw that the ocean floor itself supplied the evidence, in those mysterious magnetic stripes.

In September 1963, Vine and a senior Cambridge colleague, Drummond Matthews, published a short paper that inaugurated a new era in the earth sciences. They analysed the magnetic stripes found in the Indian Ocean the year before, by an expedition of which Matthews was chief scientist. Vine and Matthews demonstrated that the stripes could be fully accounted for, if rocks of the ocean floor were magnetised in broad bands and if every alternate band were magnetised in the 'reverse' direction – that is to say, in a sense opposite to the present magnetic field of the Earth. Vine and Matthews supposed, as Hess had not done, that the ocean floor was partly molten at the time of its formation at the mid-ocean ridge. As it cooled the rocks became permanently magnetised by the Earth's prevailing field – which was sometimes 'flipped over' from its present direction.

What this reasoning, backed by calculation, led to was an extraordinary fact of nature. The ocean floor is a huge tape recorder, telling the story of its own formation and growth. In the words of Vine and Matthews: 'If spreading of the ocean floor occurs, blocks of alternately normal and reversely magnetised material would drift away from the centre of the ridge and lie parallel to the crest of it.' Thus the ever-flipping magnetic field of the Earth imposes a series of time markers at the formation of each new segment of the ocean floor.

In 1963, the evidence on which this new theory was based was sketchy and some earth scientists even denied that the magnetic field could reverse at all. In the mid-1960s, as more and more information about the magnetic stripes became available, the Vine-Matthews theory was triumphantly successful. Allan Cox and his colleagues at the US Geological Survey were able to prove, from rocks on land, a clear-cut pattern of reversals of the Earth's magnetic field, over many millions of years. One of the most striking consequences of the mechanism invoked by Vine and Matthews is that, if the ocean floor spreads evenly on either side of the ridge, the pattern of stripes is symmetrical. The pattern on one side should be like the pattern on the other side, seen in a mirror. Careful surveys of the magnetic stripes astride the mid-ocean ridges demonstrated that this was so.

The stripes afforded the best evidence for the growth of the oceans and hence for continental drift. Their utility was even greater for they supplied a detailed history of oceanic growth and motions, and allowed the rates of growth to be compared in the different ocean basins.

Since 1969, the American drilling ship *Glomar Challenger*, in a succession of cruises, has amply confirmed the conclusions drawn from the magnetic stripes, by checking the ages of the fossils that first accumulated on each portion of the ocean floor, after its formation. The ocean floor turns out to be like a gargantuan version of the column of rocks of different ages that can be seen in a cliff face on land, only laid on its side with the youngest rocks at the mid-ocean ridge and the oldest rocks thousands of miles away towards the margins. Over the older areas of the ocean, drilling is the only way to find the age and the spreading pattern; there are no magnetic stripes, either because there were very few reversals of the magnetic field during that era of the Earth's

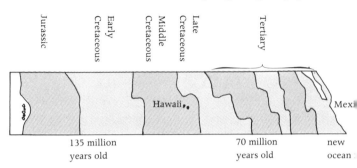

A geological column on its side. The banding of rocks of different ages, such as might be seen in a depth of a few thousand feet on land, is laid across 7000 miles of the growing Pacific floor.

Jurassic — Early Cretaceous — Middle Cretaceous — Late Cretaceous — Tertiary

Hawaii — Mex

135 million years old — 70 million years old — new ocean

history, or because the magnetic record was erased.

Why the ocean floor, at least initially, so efficient a tape-recorder is emerging from investigations of the Mid-Atlantic Ridge with the Canadian research ship *Hudson,* under the leadership of Fred Aumento of Dalhousie University. Basalt rock recovered by drilling from the ridge turns out to have very fine grains, which enable it to magnetise strongly. The molten rock has been cooled very quickly in its encounter with sea water. Quite a thin layer of rock is sufficient to account for the effects on the magnetic field observed at the surface. But the record impressed on the rocks at the time of their formation is not entirely permanent; chemical changes cause the magnetism to fade gradually, over tens of millions of years.

Converting a profession

Who first thought that the continents might have moved about the face of the globe? When the earliest realistic maps of the whole world became available, many people may have glimpsed similarities in coastlines, only to dismiss them from their minds. Francis Bacon is often credited with the idea of continental drift, but N. A. Rupke of Princeton University recently checked and found that he wrote nothing of the kind. The oldest unambiguous statement is by Theodor Lilienthal, a German theologian.

In a book published in 1756, Lilienthal interpreted *Genesis,* 10, 25 ('The name of one was Peleg; for in his days was the Earth divided') to mean a break-up of the land after Noah's Flood. He wrote that this interpretation was made likely by the fact that 'the facing coasts of many countries, though separated by the sea, have a congruent shape, so that they would almost fill one another if they stood side by side; for example, the southern parts of America and Africa.'

Combining the Biblical and geological points of view, Antonio Snider in 1858 published a book containing a map with the Atlantic closed and the continents sketchily fitted together. He, too, said that the continental mass fractured at the time of Noah's Flood but he shrewdly cited geological evidence as well, including the similarities of the fossils in American and European coal.

Continental break-up, as the result of various catastrophes (usually involving the Moon) or of a general expansion of the Earth, cropped up repeatedly in writings in the latter half of the nineteenth century and the early years of this century. But the undying credit for showing that continental drift would clear up many puzzles in the earth sciences must go to Alfred Wegener, who worked successively at the universities of Marburg, Hamburg, and Graz.

As a meteorologist, as well as a balloonist and polar explorer, Wegener belonged, as Sir Edward Bullard puts it, to 'the wrong trade union', and that was certainly one of the reasons why geologists remained unpersuaded for so long. He also mixed bad arguments in with the good ones, which made him easy game for critics. Nevertheless, in the successive editions of his slender book, *The Origin of Continents and Oceans,* published between 1915 and 1929, he refined and extended his reasoning, aided by allies and critics alike. Wegener was a generalist, one of those rare men who do not fear to learn and use branches of knowledge in which they are not formally trained, in order to arrive at a greater synthesis: conversely, over-specialisation among Wegener's opponents blinkered their imagination.

Wegener set out interlocking evidence about geological formations, fossils and past climates, showing the break-up of Pangaea, a supercontinent very like the one envisaged in current research. Although he

An early vision (1) of the pre-drift assembly of continents by Antonio Snider (1855). Alfred Wegener's reassembly of Pangaea (2) differs from the modern reconstruction (3) chiefly in that Wegener's continents were 'rubbery', while nowadays they are regarded as essentially rigid, so that Pangaea has a wide V between Asia and Africa.

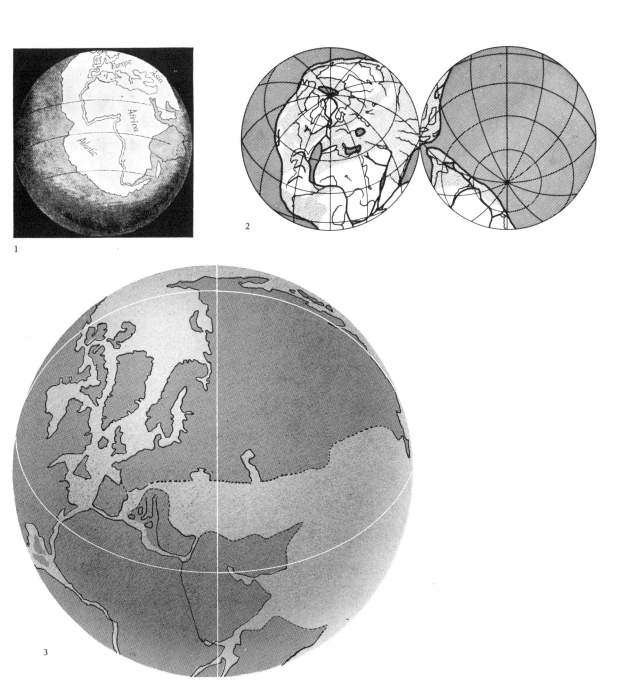

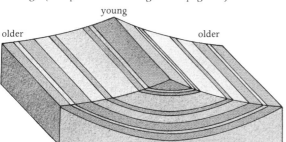

A sceptic's view of the magnetic stripes on the ocean floor. The pattern might be generated without any spreading of the floor, by layers of rock of increasing age protruding on either side of the ridge. (Compare with the diagram on page 41.)

The evenly-spaced granite domes (right) of south-west England were produced by hot, light rock welling up from a great depth and have been laid bare by the subsequent erosion of other rocks that covered them. (Drawing after A. Holmes.)

was quite wrong about the speed of the break-up (he thought that the Atlantic widened more than ten times faster than it really did) his evidence was cumulatively very impressive. He saw that the Alps and Himalayas had been made by continental collisions. Yet, by 1930, when he died on the Greenland icecap, many of the world's geologists had already dismissed continental drift as being cranky.

Recapitulating and extending the recent history of the idea of continental drift, we can see that the proof depended on information that was not available to Wegener himself: particularly the evidence gathered from the fossil magnetism of continental rocks in the 1950s, and the magnetic stripes of the ocean floor, elucidated in the early 1960s. But the debate about continental drift was just a prelude to the emergence of the even more radical idea of plate tectonics. Ingredients of it existed in Hess's notion of ocean-floor spreading and also in the assumption made by Bullard and his colleagues at Cambridge, when making fits of the continental outlines in the manner of Wegener, that the continents remained essentially rigid despite their motions.

The first to use the term plates in its present comprehensive sense was J. Tuzo Wilson of Toronto. In a now-classic paper, written in 1965, he argued that the regions where movements of the Earth's crust were concentrated were themselves connected 'into a continuous network of mobile belts about the Earth which divide the surface into several large rigid plates'.

The more precise formulation of the theory of plate movements fell to two young men working independently: Dan McKenzie of Cambridge University (assisted by Robert Parker) and Jason Morgan of Princeton University. In 1967–8 they combined geometry and geophysics to show how the shapes and motions of

plates could be determined, at least in the oceans, and how they travelled predictably as rigid units rotating about the globe in accordance with strict geometrical laws. Their ideas were quickly followed by a torrent of evidence and supporting argument, notably from studies of earthquakes by Lynn Sykes and his colleagues at the Lamont-Doherty Geological Observatory.

Anyone who doubts that man is by nature a scientific animal should consider how lightly we wear our learning and how quickly we forget what ignorance was like. Who can put himself in the frame of mind of ancestors, as reasonable as ourselves, thinking that the Sun ran around the Earth and that human beings had no kinship with the other animals? Just a decade ago, the rocky platforms on which we live our lives seemed to many quite immobile; even so recent a belief is difficult to recapture now. It may be especially hard for those who live in Britain or the countries of the southern hemisphere, where the idea of continental drift never ceased to be a plausible alternative view of the Earth; it was kept alive by Arthur Holmes and Alexander du Toit and Warren Carey, so that one is left with the feeling of having known about it all along. Life was harsher for 'drifters' in the USA where, even in the early 1960s, reputable scientists were openly laughed at for uttering what is now standard doctrine.

The conversion of a profession did not occur overnight, although for the individual earth scientist there was usually some 'last straw' of evidence after which he could never again think of the planet in the old way. For example, in Britain, Sir Edward Bullard was persuaded in the 1950s by the work of Ted Irving, who had deduced from the magnetisation of rocks that Australia must have moved through at least 40 degrees of latitude. The state of play among the younger scientists in America is well illustrated by Kenneth Hsü's account of how, in 1968, he had declared that the

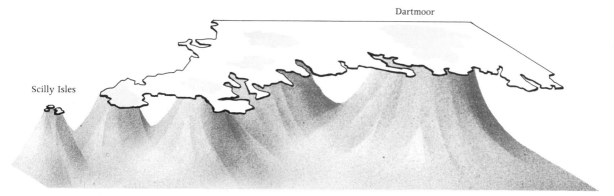

Scilly Isles

Dartmoor

'Vine story' was just a fad. 'I'm going to make you eat your words', a fellow-scientist replied, and he persuaded Hsü to join a cruise in the drilling ship *Glomar Challenger* across the South Atlantic. Hsü was converted at sea, as he saw the ocean-floor samples bearing out, hole by hole, the predictions of ocean-floor spreading. It would be invidious to tell just how slow some of the most famous earth scientists of our time were, in yielding to the new discoveries and concepts.

The unconverted

For a leading Soviet earth scientist, V. V. Beloussov, ocean-floor spreading is 'a hasty generalisation of certain data whose significance has been monstrously overestimated'. At first, Beloussov dismissed the magnetic stripes on the ocean floor as 'rather irregular scattered patches'; by 1970 he was relying on another explanation for them, as shown in the diagram opposite. Beloussov himself is a great advocate of vertical rather than horizontal movements, even to the point of sinking continents to make oceans.

Most of Beloussov's younger associates have been readier to accept the new discoveries. We were in Tokyo when Gleb Udintsev of the Moscow Institute of Oceanology came ashore from the research ship *Vityaz* and told Japanese reporters of a struggle between different points of view in the USSR, as in other countries. 'The development of science is going by the usual way', Udintsev remarked. An international geophysics congress in Moscow in 1971 helped to bring young earth scientists of the Soviet Union into contact with their western colleagues, to discuss the new theories.

Beloussov and the Russian old guard are by no means the only distinguished opponents of the idea of large-scale horizontal movements of the Earth's surface. Among others is the doyen of British geophysicists, Sir Harold Jeffreys. He argues that, if the Earth were plastic enough to allow the continents to move, then known irregularities in the Earth would iron themselves out much more rapidly than they seem able to do.

At a geological laboratory in Sweden, Hans Ramberg sustains his dissent with the aid of elaborate models: he spins mountain ranges out of putty. Ramberg, who is a Norwegian-born geologist now at Uppsala, does not challenge the recent break-up of the continents, in fact he makes models of that, too. But he thinks it has little to do with mountain-building, and, in other respects, his ideas are in line with Beloussov's. Ramberg emphasises the importance of large up or down flows of materials, driven by gravity. He deals especially with the movements that occur when heavy rock lies over lighter rock.

Ask Ramberg for a model of south-west England and he will put a layer of syrup on top of a layer of oil, in a narrow glass tank. As you watch, the oil, being less dense, forces itself up through the syrup. It does so in a predictable way. A waviness appears in the interface between the oil and syrup and then the oil rises from the crests in a row of domes, quite evenly spaced. If the oil represents granite, light rock made by heat deep below the surface of south-west England, then the rising oil makes a convincing model of the row of six granite domes that runs from Dartmoor to the Scilly Isles.

A working model of a chain of folded mountains such as the Alps needs more elaborate preparation in the laboratory at Uppsala. A layer-cake is made from wax modelling clay and various putties of different densities; differences in colour make the movements easy to follow. Some light material goes at the bottom of the cake, in order to create the instability that will

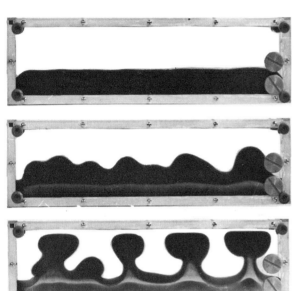

Stages in the formation of model 'domes', like the granite domes of south-west England (see previous page). The model, by Hans Ramberg of Uppsala, consists of oil rising through a denser layer of syrup. The surviving stumps in England represent the 'stalks' of the domes in the third photograph.

A cross-section of a model mountain range made by Ramberg. The model consists of layers of silicone and putty of different densities, which have been spun in a centrifuge to accelerate the action of gravity. According to Ramberg, many effects now attributed to continental movements may alternatively have been caused by vertical movements driven by gravity. Notice how the action of upwelling 'rock' layers has produced horizontal effects such as crumpling, sliding and overthrusting.

make the model change. With these sticky materials, one would have to wait a very long time for the changes to occur by the ordinary action of gravity, so Ramberg uses a centrifuge to whirl the model at high speed and thus drive it by the equivalent of 1000 to 3000 times the normal effect of gravity.

The mountain-building then happens in a matter of minutes. When the cake is removed from the machine and cut open, it shows intricate patterns of upheaval and folding that are certainly reminiscent of the cross-sections of real mountain chains. Ramberg reasons that mountain chains arise when heavy rock has spilled over lighter rock and the layers then try to change places.

From those who hold that folded mountain chains are made by continental collision, the criticism of Ramberg's modelling is that it does not show any overall narrowing of the zone of mountain-building, for which they think there is good evidence in the Alps and elsewhere. Nevertheless, Ramberg's experiments are salutary. Horizontal movements of continents may produce vertical movements of the Earth's surface, by crumpling, but vertical movements can also cause horizontal movements by toppling and sliding. Large-scale vertical movements certainly occur; indeed plate movements would be impossible without them. So the controversy is in part a 'chicken-or-egg' argument about what is cause and what is effect.

Opposition to the present triumph of the 'drifters' ought not to be sneered at. Yesterday's rebel is tomorrow's tyrant. The inevitable errors or over-simplifications of the plate theory will come to light only if some earth scientists remain as eccentric in their views as the 'drifters' were a generation ago. As one geologist remarked to me: 'Ten years ago, you couldn't become professor at a US university if you believed in continental drift; now the opposite's true.'

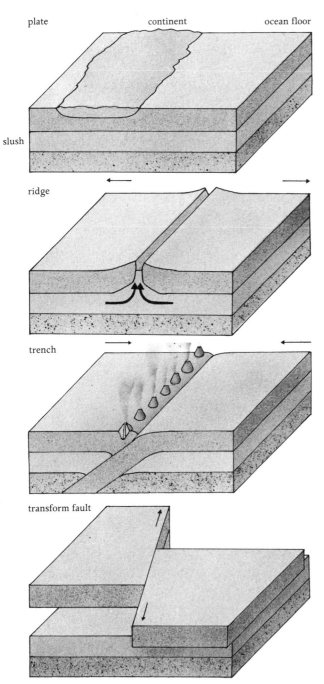

plate   continent   ocean floor

slush

ridge

trench

transform fault

Tectonics in a nutshell

Tectonics means construction – it comes from the same root as architect. Plate tectonics therefore means the construction of the Earth's geological features by the action of plates. The main ideas of plate tectonics are simple, yet they comprehend all the mechanisms that reshape the Earth's surface, except only for the details of chemical changes and of weathering. In summing them up I repeat, for the sake of clarity, some points already made.

The rocky surface and the outermost forty miles or so of the Earth consists almost entirely of cool rock that is well enough frozen to be rigid in ordinary circumstances, like the hard shell of a nut. The shell is broken into a number of plates. Heat and agitation from inside the Earth keep the plates moving about and jostling one another.

The cardinal rule of plate tectonics is that very little change occurs in the middle of a rigid plate; reconstruction is nearly all done at the edges of the plates. These edges are found, in practice, by looking at the map to see where earthquakes occur most frequently.

No gap can exist between the plates – there are no forty-mile-deep chasms in the Earth's surface. One plate can ease itself away from another but in that case hot rock instantly rises from below to fill the incipient gap. This is the situation at mid-ocean ridges and, on the continents, in rift valleys.

Nor can plates overlap to any great extent. If plates are moving towards each other one of them dips and the material of that plate passes under the edge of the other, to re-enter the Earth's interior at quite a steep angle. The bend in the downgoing plate creates an ocean trench. The sinking plate causes deep-seated earthquakes and the friction that it generates as it drives down into the Earth causes volcanoes on the far side of the trench.

The Nile Delta from a manned spacecraft. In the picture are two major plates, the African plate in the foreground and the Arabian plate in the distance. Between them is the minor Sinai plate, meeting Africa along the Gulf of Suez and Arabia along the Gulf of Aqaba. These gulfs are arms of the Red Sea, a new ocean. The Mediterranean, left, is a dying ocean.

So far, we have two kinds of plate boundaries: mid-ocean *ridges* where two plates are being manufactured at the same time, and ocean *trenches* where one plate is being swallowed up under the margin of another. There is a third kind of edge, where two plates are slipping alongside one another, without either parting or approaching to any great extent. Plate edges of this type are called *transform faults* and their identification by J. Tuzo Wilson in 1965 was one of the crucial preliminaries in the invention of plate tectonics.

Transform faults complete the outlines of the plates by joining the portions of ridges and trenches together. Wilson called them transform faults because they occur where a ridge or trench comes abruptly to an end and is transformed into something else – the fault. Transform faults are much involved in forming a new ocean when the continents split along an irregular line.

Each plate of the Earth's outer shell is thus marked out by a ring of earthquakes, following a crooked outline around ridges, faults and trenches. Some plates are very large, others are quite small, but they all interlock. Activity at one place can have effects many thousands of miles away. The growth of the plates of the Atlantic, for example, could not proceed if the plates of the Pacific were not being swallowed.

The first achievement of plate tectonics has been to explain the meaning of the main features of the oceans in terms of ridges, trenches, faults and plate motions. Now interest has switched to the continents, to see how the mechanisms of plate tectonics may account for the more complex features on land.

The continents ride as passengers on the plates. Their most important characteristic is that they cannot sink, whatever happens to the plates they are travelling on. The continental rock is not heavy enough to submerge in the dense rock of the Earth's interior. That is why the continents are much older than the oceans and

why, when continents collide, they make mountain chains instead of ocean trenches. And, because continents never 'die', they carry the scars of events long since past, in the form of old faults and lines of weakness, which can easily be reopened. Where modern plate boundaries cross continental regions, they become fuzzy: the earthquake zones are much wider than at oceanic plate boundaries.

An individual plate is by no means a permanent feature of the Earth's outer shell. A plate without continents can entirely disappear down a trench. Short of that, a plate can change shape, either by breaking along a new line or by welding itself on to another plate. Changes are especially likely wherever three plates meet. As each of them is moving in a different way, the triple junction, as it is called, will often tend to shift its position.

Infernal heat

To lose your innocence about the planet Earth you have to see an active volcano. Knowing by hearsay or by the long-cold evidence of ancient lavas that the Earth's surface has originated in molten rock welling up from the interior is one thing. It is something else to stand amazed at the crater's edge bathed in steam, choking in the sulphur with a rain of pumice on head and shoulders, watching it happen.

'It sounds as if there should be something to see', Donald Peterson remarked as we crunched our way up a slope of frozen lava that was like black meringue. Peterson, head of the Hawaiian Volcano Observatory, took us to Mauna Ulu, the 'growing mountain', which had burst out among the ferns and ohia trees on the flank of Kilauea, in 1969. Two years later a cone stood 300 feet above the jungle: a very modest little volcano and approachable, with caution, through the screens

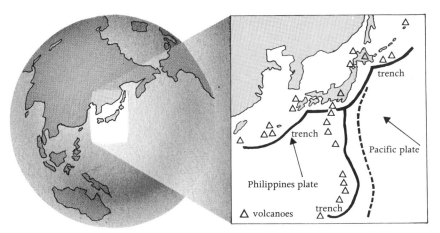

The ocean trenches south of Japan. The T-junction of plate boundaries tends to shift and the south-pointing trench (Bonin trench) seems to have migrated westwards from the dotted position.

laid across the foamy landscape by steam vents. The portents were thin blue smoke from the summit and a roaring noise like ocean breakers that grew louder as we climbed. A day or two before, part of the crater rim had fallen away and a great crack in the ground indicated that our vantage point, too, was due to go.

A yellow-hot fountain of molten rock was spurting out of the lake of lava that filled the crater. As it splashed against the far side of the crater, the fountain made the wall glow red. The contents of the lake, crusty plates of lava laced with glowing cracks, flowed into the fountain which regurgitated them. This crater had once fountained to a height of 1800 feet; that day the fountain rose a mere 100 feet but it was quite impressive enough for me as I stood wondering whether the ground shook, or only myself, and expecting momentarily to be turned into a pillar of basalt.

As we watched, another fountain appeared capriciously; one had been terrifying but two seemed playful, in spite of the additional radiated heat and the warm pumice that now began to fall on us. Even with this little volcano, one had a clear sense of how the Earth had built the island high above the floor of the Pacific. Previously we had watched lava from this same eruption sizzling slowly into the sea, adding new acres to the county of Hawaii; the pit was creative rather than satanic. But then the ground really did shake, the crack behind us widened noisily and we were not slow to follow Peterson's suggestion that we might retreat.

The material coming up at Hawaii is tapped from a zone of partially molten rock that forms a layer encircling the Earth, beneath the plates. The deep structure of the Earth, and also conflicting theories about Hawaii itself, will be outlined in Chapter 4. For present purposes our little volcano represents the raw material, principally basalt, which plate movements will reprocess over hundreds of millions of years, to make the more familiar rocks of continents. More simply, it denotes the infernal heat of the Earth.

Deep mines are warm and the core of our planet is molten. The Earth has its own heating system in the form of small amounts of radioactive elements (chiefly uranium, thorium and potassium) scattered in all rocks. Heat comes out of the ground, not only where there are volcanoes or hot springs but everywhere, through the bed of the ocean as well as through mountains. At the surface, the flow of terrestrial heat is very small compared with the heat supplied by the Sun's rays; it would take two months to melt a millimetre of ice. Even so, its cumulative effects are enormous. All plate movements, volcanic eruptions and the making of new ocean floors depend upon this infernal heat, and so does the creation of the Earth's magnetic field.

Continental rocks contain much higher concentrations of radioactive elements than does either the floor of the oceans or the deep interior of the Earth. Yet the average heat flow through the ocean floor turned out to be just about the same as the average flow through the surface of the land. Elaborate arguments developed in the 1950s about why this should be so and it was even offered as evidence against the possibility of continental drift.

Nowadays, most of the oceanic heat flow can be explained as a by-product of the growth of the plates at the mid-ocean ridges. The difficulty fades: the similarity of heat flow in continents and oceans is simply a coincidence, with two very different processes giving the same answer.

Contrasts in heat flow are now more interesting than the similarities. The memory of past heating persists for a hundred million years or more. For example, the deep-lying rocks of western Scotland are still warm from the opening of the Atlantic ocean. The

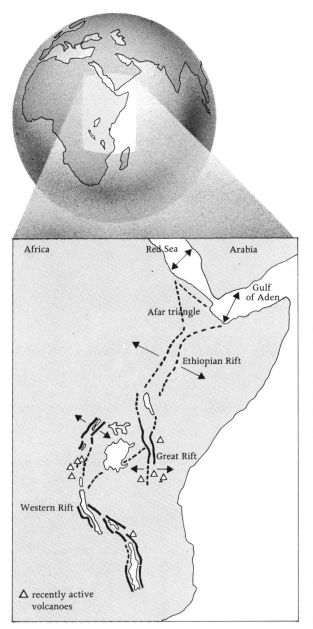

Africa   Red Sea   Arabia

Gulf
of Aden

Afar triangle

Ethiopian Rift

Great Rift

Western Rift

△ recently active
    volcanoes

A slight divergence in the direction of opening of the Red Sea and the Gulf of Aden has been associated with the remarkable rifting of the African continent. The Afar triangle is a piece of the Red Sea basin that has become incorporated into Africa.

heat flow from the ocean floor is highest at the mid-ocean ridges. It falls off as one travels away from the ridges and the ocean floor become progressively older.

In the continents, too, there is a falling off with age, the heat flow being high in the young mountain-belts and lowest in the oldest parts of the continental surfaces. The continents are very old compared with the oceans and, in places, they have begun to run out of radio-activity. Rocks 3000 million years old have by now spent much of their endowment of radioactivity – more than one-third of the uranium and three-quarters of the potassium.

The commonest types of behaviour at plate boundaries have already been mentioned:
1. rifting – two plates moving apart.
2. transform faulting – two plates sidling past each other.
3. destruction of ocean floor – two plates moving together.

In various parts of the world we can see examples of these processes going on now, and how the new geology accounts for them. In the oceans, the events proceed rather simply; on land, they become more complicated and therefore more interesting. The examples that follow are of activities at plate boundaries that are visible on land.

Rifting in East Africa and Iceland

Drive a little way west of Nairobi, the capital of Kenya, and you find that the ground suddenly drops away; you are on a cliff running north and south as far as the eye can see and before you is the Great Rift Valley of East Africa. Better still, fly over the valley and see it for what it is, a long rip in the Earth's surface, streaked with rocky terraces and punctured by the craters of

volcanoes. On either side, up to forty miles apart, imposing scarps face each other, showing where the ground has subsided between them. The full extent of the fall is not apparent because volcanic outpourings have largely filled the valley, to a depth of more than a mile. Even while the volcanoes are quiet, hot springs and earthquakes testify to continuing activity.

The valley is part of a great system of rifts that stretches south to Mozambique and north through Ethiopia to the Red Sea, with a western branch running along the Congo border. Many have been tempted to say that these rifts mark an early stage in a break-up of Africa and that the eastern part of the continent is about to go voyaging. That both overstates and oversimplifies

the situation. Africa has been tearing itself apart along the line of the rifts but very slowly indeed. If it was ever promising to make a new ocean where the game of Kenya and Tanzania graze, it failed.

In twenty million years, the land on the two sides of the Great Rift Valley has eased apart by no more than about six miles. In the same period the Red Sea, a successful young ocean, made by the severance of Arabia from Africa, has widened by 200 miles. Faster-moving plates elsewhere in the world have travelled much greater distances. Before the discovery of those plate motions, the African rift seemed to be a very remarkable feature, telling of horizontal movements in the Earth's surface. Today we can see it as a minor

53

How the North Atlantic Ocean would appear if one could see it drained of water. The topography is simplified to emphasise the main features. Note the high 'cliffs' of the continental margins; it was these that fitted together before the opening of the ocean.

The mid-ocean ridge dominates the ocean basin and transform faults cut across it. Iceland (top right) is a part of the mid-ocean ridge that has broken the surface. The photograph shows the new volcanic island of Surtsey, off Iceland.

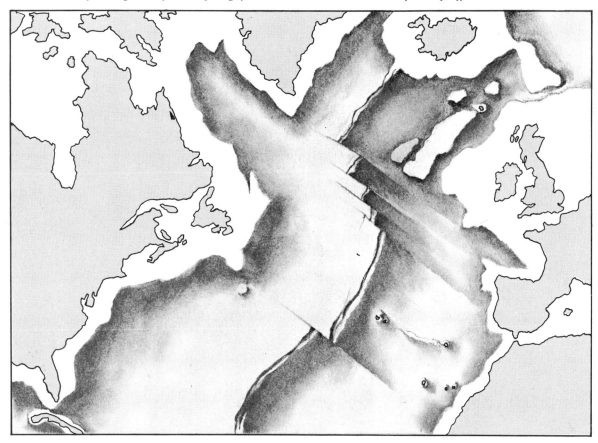

adjustment to more pronounced movements that opened the new oceans in the Red Sea and the Gulf of Aden.

The African rift nevertheless gives an impression – possibly misleading in some respects – of what the torn edges of the present continents may have been like during the early stages of continental break-up. The rifting in Africa has occurred along lines of weakness in the old fabric of the continent, left by mountain-building events in the distant past. In the most spectacular stretches of the rift, those near Nairobi and in Ethiopa, the continent heaved and made huge domes before the rifting began. If this also happens during a major continental break-up, it is evidently followed by a sagging of the edges as the new ocean widens. The zone of uplift remains in the middle of the ocean, as in the Atlantic today.

Iceland is a young addition to the living space on Earth. Both its newness and one of its modes of creation were well illustrated in 1963, when the cook

of a fishing boat saw the sea boiling off the coast and within twenty-four hours the territory of Iceland was extended by the new volcanic island of Surtsey. Iceland is made by the Earth bleeding through the incipient crack between the American and Eurasian plates. It is one of the few visible parts of a huge wound that is mostly hidden underwater – the Mid-Atlantic Ridge.

If you could see the Atlantic drained of its water, the scene would be dominated by the great chain of underwater mountains that rise as high above the ocean basins as the Alps or the Rockies stand above sea-level. It zigzags exactly down the middle of the ocean. On either side of the ocean basins high cliffs stand, the edges of the continental shelves where the land was torn apart to make the Atlantic. But the ridge is where the action is. It marks the boundary between the plates and also their region of growth, and in Iceland you stand on it.

In continuation of the deep groove that runs all along the underwater crest of the Mid-Atlantic Ridge, Iceland

The San Andreas fault near San Francisco. Airborne radar exaggerates the relief and makes the fault stand out clearly. The San Andreas is the lowermost dark streak and there are other faults running beside it. San Francisco itself is just off the picture to the right but the cross on the Bay coast is San Francisco Airport. Just inland of the airport is San Andreas Lake, from which the fault takes its name.

has its own rift valleys. These are the production lines, both for the island itself and for the huge Eurasian and American plates. Iceland, like the rest of the ridge, grows from the middle outwards. Wedges of hot rock come up along the central lines of the rift valleys, to fill the cracks that form as the plates move apart. Gashes in the ground tell of the tension. The volcanoes and the free hot water that accompany the process form part of the Icelander's way of life.

The growth of Iceland can now be detected with accurate surveying equipment. A group led by Ronald Mason of Imperial College, London, has set up widely spaced pillars at the Iceland rifts. Periodically a team visits them with an optical-cum-electronic system which employs a beam of light to measure the distances between the pillars to an accuracy of a few thousandths of an inch. From year to year, the pillars move perceptibly apart, carried on the widening territory of Iceland.

The rates of movement are about a quarter of an inch a year. They are not incompatible with other Icelandic measurements made by an American group, using a different technique. The London group has some evidence that, in south-west Iceland, the rifting is accompanied by a leftwards slipping motion. So it seems that, at least locally and temporarily, the Mid-Atlantic Ridge may also be acting like a transform fault.

A transform fault in California

The notorious San Andreas fault runs through California and makes houses expensive to insure in that state. It takes its name from a lake on the fault-line near San Francisco and, together with adjoining networks of lesser faults, it is the source of the ground movements that cause the earthquakes of California. In the new language of plate theory, the San Andreas is a trans-

The transform faults of the world, where plates sidle past one another.

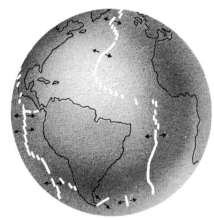

The ocean ridges of the world, where plates are moving apart and growing.

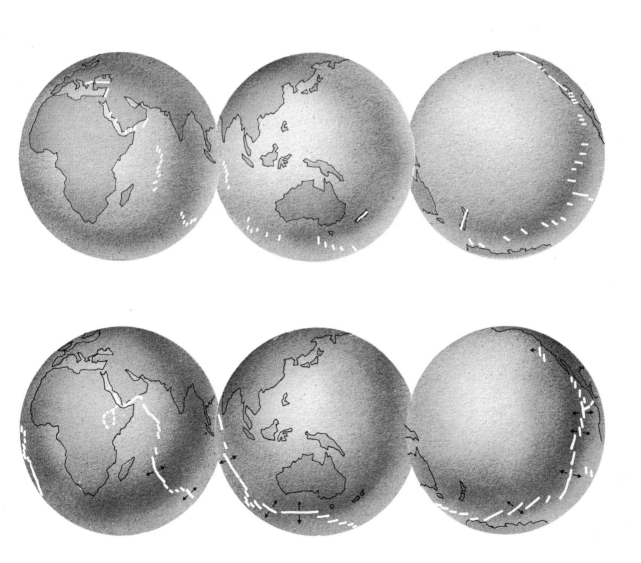

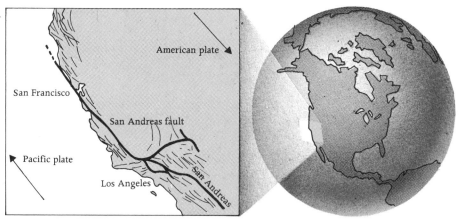

The San Andreas fault of California bounding the American and Pacific plates. The continent is shattered in a host of associated faults.

form fault – a boundary where two huge plates of the Earth's crust, the American and Pacific plates, sidle past each other with tendencies, here and there, to pull apart or press together. The Pacific plate is growing from the East Pacific Rise, from a 'mid-ocean' ridge that is no longer in the middle because so much of its ocean has been destroyed by the westward motion of the Americas.

The Pacific plate (the largest plate of the Pacific Ocean) carries a small slice of the USA with it as it travels at about two and half inches a year, north-east-wards, relative to the main body of the continent. A bystander with the necessary patience, spending a lifetime on one side of the fault, would see the houses and hills opposite move about fourteen feet to the right. If present motions continue for twenty million years, Los Angeles will lie as far to the north of San Francisco as it now lies to the south. But, as Californians know to their cost, the scenery does not travel smoothly; instead the fault tends to stick and then to slip suddenly, releasing the accumulated strain and sending shock waves through the surrounding land.

By rough, sometimes treacherous roads we followed a segment of the San Andreas fault for many miles. The fault is not normally a crack in the ground, or anything like that; you track it as you would an old Roman road in England. There are conspicuous stretches, in lines of lakes or offset streams; elsewhere, you just have a sense of straightness in the landscape – which is very plain in aerial photographs. That straightness is the obvious feature of the San Andreas fault. Less obvious is that the hills lying beside or athwart the San Andreas fault have been made by the same general motion as the fault itself.

California is the most highly instrumented and closely studied of all the world's earthquake regions, yet the Los Angeles (San Fernando) earthquake of 1971

took the seismologists by surprise. It occurred along fault lines that had not moved for thousands of years and it made the San Gabriel mountains to the north of Los Angeles grow a little higher. Here, as in some other places in California, the fault lines do not lie parallel to the San Andreas fault and the direction of motion; instead, they run at an angle, as if trying to bar the motion of the plates. They cannot do so, of course, and as the Pacific and American plates continue on their way, they heave up the mountains. The San Gabriel mountains would be as high as the Himalayas, and Hollywood would be like a seaside Nepal, were the mountains not worn away almost as fast as they grew.

For Tanya Atwater, this is how the American West has been made, or at least remodelled, during the pro-longed conflict between the continent and the floor of the Pacific ocean. Atwater was still a student oceano-grapher at the Scripps Institution when in 1969 she insisted, more emphatically and logically than anyone had done before, on deducing from the observed motions of the ocean floor the inevitable effects on the fabric of America.

During the previous decade, other earth scientists, including Warren Hamilton of the US Geological Survey, had perceived a kinship with the San Andreas fault in many other lines of slippage, ancient and modern, in western North America. The whole region is 'soft'. The boundary between the rigid Pacific and American plates is not confined to the San Andreas line but occupies a diffuse belt stretching a thousand miles inland.

Atwater's account endorsed this general point of view, but she was able to add more detail, especially about recent events. The spreading of the floor of the north-east Pacific, which she had been investigating with H. W. Menard, showed that even a constant relative motion between the main Pacific plate and

A road intersection smashed in the Los Angeles (San Fernando) earthquake of February 1971. The earthquake originated in a fault running at an angle to the San Andreas fault and the mountains north of Los Angeles became a little higher.

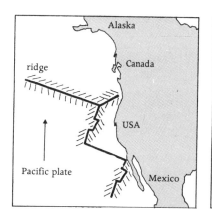

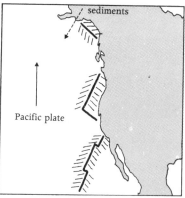

There are thick sediments on the Pacific plate south of Alaska which could not have reached it from Alaska before a certain time, depending on how far the plate has moved. (After T. Atwater.)

the American plate implied an ever-changing pattern of events in North America. These included the destruction, under the margin of the continent, of oceanic plates of which little or nothing now remains.

Volcanoes were prevalent throughout the western United States forty million years ago but died out in the south around —1 MY. They marked the consumption, in that region, of a now-vanished oceanic plate. The volcanoes that are still occasionally active in the north-west are due to the slow destruction of the last vestige of another plate.

Tugging by the Pacific motion created the curious Basin and Range province of Nevada and other western states, in which huge blocks of the Earth's surface have tilted and subsided. This occurred between —20 MY and —5 MY, across faults running obliquely into North America.

Compression by the same kind of motion occurred at various places and times, but notably in the making of Californian mountains running at an angle to the coast, during the five million years since the San Andreas fault 'jumped ashore'.

Individually, mechanisms of these kinds have occurred to other geologists; Atwater's achievement is to tie them all together in one coherent account. She assumes that the movement along the axis of the San Andreas fault was enormous, though not confined to the fault itself. During the past twenty-five million years, according to Atwater, the Pacific plate has moved several hundred miles past North America.

Some geologists remained very sceptical. One of them, Donald Scholl of the US Geological Survey, resolved to put the whole theory to a searching test. It was a venture in scientific detective-work that took *Glomar Challenger* near to the Aleutian Islands in the north Pacific. Off these islands is an ocean trench where the Pacific plate is being consumed. South of the

Death Valley, California. This is one of many valleys in the western USA formed by a tugging action of the Pacific plate upon the continent, in the so-called Basin and Range region.

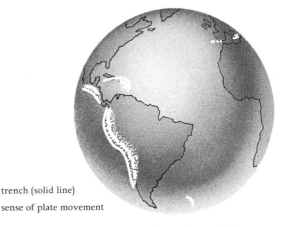

trench (solid line)

sense of plate movement

trench, the ocean floor was known to carry a thick load of hardened mud and Scholl wanted to find out where the mud came from, and when.

The most obvious source was Alaska but, if the Pacific plate had travelled at the maximum rate thought possible by Atwater, it would not have been in a position to receive mud from Alaska until about −25 MY. If the mud were older than that, as Scholl suspected it might be, and if it had nevertheless come from Alaska, then the Pacific plate must have moved more slowly.

In July 1971, the drilling was made, to recover samples of the mud from the ocean floor. In *Challenger*, the mud went under the microscope. The palaeontologist looked for characteristic fossils that, to an expert eye, would reveal the age of the mud. The answer was that Scholl's hunch seemed right – the sediments were about 50 million years old and looked like Alaskan material. At the time of writing the interpretations are being debated but the result still allows for up to 500 miles of relative motion along the San Andreas axis. Although it suggests less steady motion than Atwater had thought likely, her geological interpretations for western North America remain plausible.

Arcs of volcanic islands

At the entrance to Tokyo Bay, you can see cliffs that look like giant staircases, built by a long succession of earthquakes. The waves of the ocean eat away the rocks at the shoreline, making a little beach, but then that beach is heaved up in an earthquake and another beach begins to form. The steps are quite regular because the rocks have a natural strength and they break whenever that limit is reached. Fortunately, Japan is fairly weak, otherwise that country's earthquakes would be even worse than they are. Ground movements of about nine feet in 1923 wrecked Tokyo and Yokohama. In Alaska,

for comparison, earthquakes have been known to displace rocks by more than forty feet.

Japan owes much of its present land surface to the very processes that now plague its inhabitants. It lies alongside deep trenches in the ocean floor, where plates are being driven back into the deep interior of the Earth. The plates do not go quietly. They cause earthquakes and they throw up volcanoes – including Japan's sacred mountain, Fuji, which last erupted in the eighteenth century. The earthquakes originate quite near to the surface on the Pacific side of Japan, but on the Asian side they are much deeper. In fact, the plots of the earthquakes give ghostly outlines of the downgoing plates, sloping steeply into the Earth like a moving staircase.

The volcanoes, active and dormant, arrange themselves in gently curving arcs behind the ocean trenches. They lie at a distance where the upper surface of a plate, descending underneath them, has generated sufficient friction to heat the rocks to the point of eruption. Wherever plates are destroyed, volcanoes tend to appear.

Japan is far from being the simplest case. It has a long geological history and may have broken away from the mainland of Asia. Moreover, it lies at the junction of three plates, beside a 'T'-shaped group of trenches: the tail of the 'T' is a plate descending unusually steeply and growing a chain of new islands that stretches away to the south. Hiroo Kanamori of the Earthquake Research Institute in Tokyo thinks that here the plate has completely broken and is sinking almost vertically into the Earth's interior.

In several other 'island arcs' in the world the situation is plainer. One plate goes down at a trench at a slope of about forty-five degrees and a row of brand-new volcanic islands grows up at a distance of about 100 miles behind the trench. That has happened in,

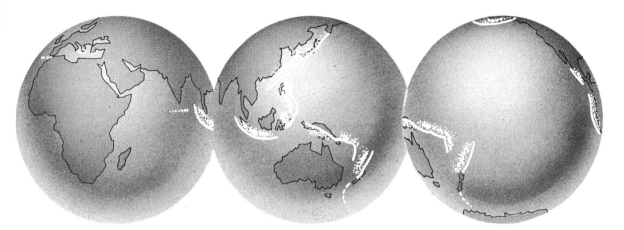

for example, the Aleutians. Island arcs of varying degrees of complexity include, in the Pacific region, the Kurile islands north of Japan, the Tonga islands far to the south, the Philippine islands and New Guinea. Indonesia is a very active island arc – remember Krakatao. A group of islands off Antarctica have a similar origin. The West Indies, the setting for the Port-Royal earthquake and the volcanic destruction of St Pierre, make up an island arc that nibbles at the floor of the Atlantic. Some Aegean islands of the Mediterranean, including Thera-Atlantis, constitute another island arc.

The possible situations alongside an ocean trench, where an oceanic plate is being destroyed, can vary greatly, from a virgin arc of volcanic islands like the Aleutians, through a volcanic microcontinent like Japan, to a volcanic edge of a major continent. The chief example of this last possibility is the west coast of South America, which is overriding an oceanic plate in the Peru-Chile trench. Its continental bulk gives no immunity against the usual consequences of plate consumption. There is much volcanic activity in the Andes; Cotopaxi in Ecuador is the highest of the world's active volcanoes. And, of course, Chile and Peru have been the scenes of some of the worst earthquakes of recent years.

A graveyard of plates

Waves from earthquakes have been recorded systematically for only about a hundred years, since the invention of modern seismographs. These are instruments that usually carry, on a boom, a heavy mass which tends to stay put when its support moves in response to a ground tremor. The instrument registers the relative motions of the support and the weight, in the well-known wavy traces.

Seismology, the study of earth movements, came into prominence in international politics in 1958 as a means of detecting underground nuclear explosions. That was when the governments of the USA, USSR and Britain were groping towards a treaty banning the testing of nuclear weapons. In the outcome, the test ban did not include underground testing. Nevertheless, the political interest, and the money for seismology that it unleashed, led to very rapid advances. What had previously been done by a small number of enthusiasts, using all sorts of different instruments, became a co-ordinated operation.

The US Coast and Geodetic Survey set up a worldwide network of more than a hundred seismograph stations equipped with standard instruments. Although nuclear explosions prompted the creation of this network, it has transformed man's ability to record and pinpoint natural earthquakes and today it is an essential source of information about plate movements.

Another line of attack on the explosion-earthquake problem was by the development of seismic instruments of unprecedented sensitivity, including the construction, in 1965, of a huge array in the state of Montana. LASA, or the 'large aperture seismic array', incorporates techniques devised in the USA and Britain for extracting the last possible jog of information out of the tremors of the Earth.

LASA consists of no fewer than 525 seismometers, buried 200 feet underground and deployed over an area as large as Wales or Massachusetts. A seismic wave coming through the Earth from a particular direction washes over the widely spaced detectors at different moments; in fact it takes about ten seconds to cross the whole array. Just as a man's brain discerns the direction of a sound by the slight differences in its time of arrival at his two ears, so LASA's electronics can tell the direction of seismic waves. LASA is like a telescope

The Great Rift Valley of Kenya. Volcanic rocks filling the floor of the valley conceal the full depth of this tear in the fabric of East Africa. When the Red Sea and the Gulf of Aden opened, *East Africa was stretched and it broke here along old lines of weakness. The extent of the rifting is very small compared with the rifting that made the Atlantic, for example.*

that can look in any chosen direction inside the Earth. The use of so many instruments in combination also gives LASA the ability to pick out waves that would otherwise be hidden by small, random agitations of the Earth. Tremors that move the ground by a millionth of an inch are easily detected by LASA.

An elaborate communications and computing system makes the information from LASA instantly available in Washington. The signals also go to the Massachusetts Institute of Technology, where the instrument was conceived. At MIT's Lincoln Laboratory, research continues into ways of discriminating between explosions and earthquakes, under the leadership of David Davies, a distinguished young seismologist, formerly of Cambridge University.

Davies himself has been particularly interested in a feature of the seismic signals that bears strongly on the history of plate motions as well as on the detection of nuclear tests. In 1969, with Dan McKenzie, he cleared up a four-year-old mystery about an American nuclear test in the Aleutian Islands, codenamed Longshot. Hundreds of seismograph stations recorded the event but, when the location of the explosion was calculated by well-proved methods, it was not where the explosion actually occurred. It seemed to lie forty miles beneath the Earth's surface!

The explanation found by Davies and McKenzie was that the waves were being temporarily trapped in the cold plate that is descending deep into the Earth under the Aleutians island arc, as the Pacific floor moves northwards. The seismographs were, in effect, seeing the descending plate, as one might see a sheet of glass in a fish tank. Since then, other descending plates have been detected inside the Earth. Besides those going down from known ocean trenches, Davies and his colleagues are finding relics of plates that disappeared from the surface long ago. They still lie as cool slabs in the Earth's interior, which is the graveyard of many ancient oceanic plates.

One of the old slabs is right under the Arctic nuclear testing ground of Novaya Zemlya and explains why seismic records of Soviet explosions there have always looked very peculiar. Davies identifies it as the remains of the old oceanic plate that was swallowed up about 250 million years ago, when Asia was running into Europe to make the Ural mountains. Davies confesses he is surprised that the plate has survived so long. The geological importance of this discovery is that it provides a bridge between 'recent' events – those occurring from the break-up of Pangaea to the present day – and similar processes going on several hundred million years ago, before Pangaea existed.

Chapter 3   How Long Adrift?

Controversy has now shifted to the question of whether continents remained fixed before the comparatively recent break-up of a supercontinent. In the light of evidence that they were moving long before that, many familiar features of our landscapes can now be explained by the resulting continental collisions.

Betts Cove in Newfoundland is off almost everyone's beaten track and at first sight it deserves to be. Miners have been there for the copper but there is no towering mountain or great river or anything else to draw the eye, except perhaps for a certain colourful diversity in the rocks. Yet it is a key site for one of the young arch-revolutionaries of the earth sciences, John Dewey, a British geologist now based at the State University of New York at Albany. Dewey and his colleague John Bird have gone again and again to Newfoundland to scramble over the rocks of Betts Cove and nearby terrain, mapping the ruins of a lost ocean.

Across the world, in Cyprus, near other copper mines, peculiar rocks reveal that island to be a piece of an ocean plate that has been crumpled above sea-level (see p. 33). The stunning feature of Betts Cove is that its assembly of rocks is almost identical with those in Cyprus, but far older. In Betts Cove, as in Cyprus, are amassed pillow-like lumps from an ocean floor; there are walls of basalt, too, close-packed, made by rock filling the cracks in a growing ocean floor; there is even dense rock from the deep interior of the Earth, that has been squeezed to the surface.

Cyprus has been newly formed by plate movements now plainly in progress, associated with the narrowing of the Mediterranean. The ocean at Betts Cove was annihilated nearly 500 million years ago. Dewey and Bird interpret the evidence, which also includes volcanic ash on the top of the former ocean floor, to mean that off the east coast of North America, around —500 MY, there lay an island arc. Subsequently, the great

In both Cyprus (upper photograph) and Newfoundland (below) sets of peculiar rocks occur among the hills, which represent a cross-section of the ocean floor. In Cyprus the rocks have been forced out of the sea quite recently by pressure between plates. The Newfoundland rocks are much older, suggesting that plates have been acting as at present for a long time.

basin separating the volcanic islands from the mainland was destroyed, leaving only a little material from its floor heaped on the land.

The implications are far-reaching. The present Atlantic Ocean has come into being only during the past 180 million years or so. Before that, North America, Europe and Africa were tight-fitted in the supercontinent of Pangaea. If they were always that way, there could have been no ocean in which an island arc might grow; no ocean floor to heap its relics on the shore at Betts Cove. There must have been a former North Atlantic ocean, with roughly the same coastlines as at present, though not necessarily so wide.

If this reasoning is correct, Pangaea, which has been our ancestral reference for the recent motions of the continents, was only a short-lived feature in the history of the Earth, assembled by chance from continents that were travelling about the globe long before. Plate tectonics is then the dominant theme for much of the Earth's history. This is indeed a sweeping deduction from a few rocks in Newfoundland. There is plenty more evidence, although the comparison of Betts Cove with Cyprus does represent one of the better pieces.

With all but a few earth scientists persuaded of the break-up of Pangaea, the main controversy has now shifted to the question of how long the continents have spent on the move. One group declares that the recent episode of continental break-up and drifting is exceptional and that Pangaea existed as one continental mass, or perhaps two, throughout most of the Earth's history. For example, Patrick Hurley of the Massachusetts Institute of Technology, who helped to confirm the rift between Africa and South America by matching rocks on the two continents, cites similar evidence *against* earlier motions. He suggests that the old rocks of Pangaea were too coherent for them to have been brought together by chance collisions.

A. E. J. Engel of the Scripps Institution takes much the same line. The continents grew steadily outwards from their oldest cores, making something like a tree-ring structure of increasingly younger rocks. This, for Engel, argues that they cannot have been adrift – although he thinks that there was a much earlier period of plate movements and drift, more than 2500 million years ago, when the cores of the continents were first being assembled out of micro-continents. These are reasoned objections to perpetual drift; but there is also a conservative rump of 'anti-drifters' who merely wish to save as much as possible of their former beliefs by confining plate tectonics to the past 200 million years.

The main current of thought, among the revolutionaries and those who acquiesce, has the continents moving for a very long time. There is most confidence about supposed major movements going back 600 million years – that is to say, throughout the period in which abundant fossil remains of living organisms have lent clarity and precision to the investigation of rocks and the events that made them. There is a fairly widespread conviction that drifting was in progress as early as −2500 MY. Some say, with Engel, that even older rocks suggest plate action. They are then, more as a matter of principle than of evidence, happy to assume that the process has continued without interruption since −3500 MY.

Others suspect a gradual change in the nature of the Earth's shell – perhaps an alteration in temperature, or the distribution of heat, perhaps a chemical change. If so, that is a good reason to hesitate before invoking plate movements earlier than −2500 MY, especially as our knowledge of that era is scanty. The next chapter will return to the earliest origins of the continents. Here we shall pursue the evidence of intervening movements, including that at Betts Cove and similar sites, and tell the story of the older mountains of the Earth.

Ice in the Sahara Desert. These scratches were made by an ice sheet that covered the central Sahara about 450 million years ago, when North Africa was near the South Pole. At that time North America was enjoying a tropical climate, so that a wide ocean must have existed between the two continents before they came together in the supercontinent of Pangaea.

Clues to past motions

We talk of vanished oceans. The clearest markers of recent movements of plates, and of the continents they carry, are the magnetic stripes on the present ocean floor. If a former Atlantic existed, its destruction also erased any tape-recorded message of its growth that it might have carried. Nor is anything to be gained by looking for oceans that opened in the wake of continents which were moving together to destroy the former Atlantic. Virtually all the oceans have renewed themselves within the past 200 million years. We have to turn to other clues to earlier continental drift.

One is the fossil magnetism of old rocks within the continents. As recounted in the preceding chapter, investigations of rock magnetism on land had provided evidence for recent continental drift, several years before the decisive interpretation of the magnetic patterns detected at sea. For rock magnetism, as for any other traces of events on the planet, difficulties and misgivings become greater the farther back in time one looks. The right kind of material is harder to find and the chances grow that the record has been smudged by subsequent events. Nevertheless, as early as 1960 the London group of rock magnetists spoke confidently of movements of Europe, North America and Australia continuing uninterruptedly since −500 MY. The London group is now disbanded, but the work has continued in several other centres. Curiously enough, the controversy about drift versus polar wandering (pp. 38–9) has been renewed for the older rocks. K. M. Creer of Newcastle contends that before the break-up of Pangaea the Earth's poles may have shifted in relation to the land, but not the continents in relation to one another. Others, including Ted Irving (Ottawa), Michael McElhinny (Canberra) and James Briden (Leeds), are satisfied that the continents did change their relative positions, before −200 MY.

Basalt dykes in Montana. They were made by hot rock welling up to fill long, narrow cracks in the Earth's surface. Dykes are a sign of tension, between plates that have moved apart or within a continent that has been stretched by plate action.

The present speed record for continental movements is held by Gondwanaland, the cluster of southern continents. Around —300 MY, it travelled 3000 miles in 20 million years, as determined by Irving in Australian rocks. That is about ten times as fast as the present motion between Europe and North America, which is opening the Atlantic. Because the raw figures from laboratory measurements on rocks leave only limited scope for interpretation according to the fancies of the investigator, rock magnetism is potentially the most objective guide to earlier continental movements.

Also embedded in the rocks is a further reasonably unequivocal source of information – signs of climates prevailing in the distant past. The simplest and most striking example concerns the relative positions of the eastern USA and North Africa. In the time of Pangaea, they lay side by side. But, back in the period —500 to —440 MY, there are plain signs that much of the present Sahara desert was covered in an ice sheet, including Morocco which, in Pangaea, was butted against North America. The implication is that North Africa was near the South Pole. If Pangaea was intact at that time, you would expect to find symptoms of a rigorous climate in the rocks of the same age in eastern North America. Not so; the marine fossils there testify to a warm, subtropical climate and North Africa and North America must have been separated by an ocean several thousand miles wide. This is a conclusion that fits well with the results of rock magnetism. An important point to note is that the climatic and magnetic evidence gives information about the distance, north or south, from the poles or the equator; they say little or nothing about the distribution or movements of continents east and west around the world.

The study of the rocks at Betts Cove, with which this chapter opened, illustrates a third approach which may, in the long run, be the most instructive about plate tectonics before —200 MY. The idea is to see what kinds of rocks and structures different kinds of plate boundaries are producing today, as a result of known plate movements, and then to look in older mountain chains for the same kinds of rocks and structures. These older formations may then be explained by plate movements and continental drift. This exploration of the remoter past calls for a more systematic consideration of the geological clues left by the three kinds of plate activity: growth at the mid-ocean ridges, tearing and sidling along transform faults, and destruction of the plates in ocean trenches.

Unfortunately, if one is looking for direct evidence of past growth of plates, the mid-ocean ridges themselves do not survive. They wither when they become inactive and cool, and they can be destroyed along with the rest of their oceanic plates; this has happened under the west of North America, where the East Pacific Rise has partly descended under the margin of the continent.

The opening of new oceans in the past also involved the splitting of a continent, in a flurry of volcanic activity along a rift valley. The evidence of this event persists plainly for a while along the virgin continental margins, as in the Atlantic today. But sooner or later such margins become involved in the destruction of the adjacent ocean floor, which radically alters the margins. Something of the chemical signature of rifting, which is essentially basalt rock, may persist despite the later events, either as material erupted from volcanoes or as wedges of basalt that filled cracks in the stretched surface of the splitting continent.

Incipient plate growth may, on the other hand, be recorded in fossil form, if all that happened was the production of a rift valley, like those in East Africa Ancient rifts tend to fill and become buried, but geological detective work or soundings by shock waves can reveal them.

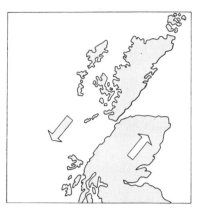

An ancient transform fault in Scotland. About 350 million years ago, a plate boundary cut across the country and shifted the two parts to their modern positions. Erosion along the fault has produced the Great Glen, below.

Transform faults on land, like the present-day San Andreas, leave a fossil record in the form of a mismatch of rocks and structures on either side of a long, straight fault-line. There are many features of the Earth which are roughly straight, including quite a few mountain chains, but only a transform fault is capable of producing a line of geometrical perfection. Strictly speaking, it is not straight; it is an arc of a very large circle. Transform faults are highly vulnerable to erosion and, with the passage of time, perfection may crumble; there persists what I described in the San Andreas as 'sense of straightness'.

A transform fault of considerable age occurs in Scotland, making the Great Glen and Loch Ness. It was active about 350 million years ago and it altered the outline of Britain by shifting the northernmost part sixty-five miles to the south-west (see the maps). It allows us to say there was a plate boundary cutting through the continental material at that time, and that the sense of relative motion was as indicated. The extent of the motion is not particularly impressive; had it been much greater, the north of Britain would have been wrenched right off. The fault is not quite inactive; minor earthquakes occur, on average, every three or four years, to rouse the long-sought monster of Loch Ness.

The remaining type of plate boundary, for vestiges of which the geologists can now search in old structures, is the zone of plate destruction in ocean trenches. This kind of zone makes durable mountains and is the most important for retelling the story of the Earth.

The tell-tale mountains

The boldest notion of the new geology is very simple. It is that every great mountain chain on the planet has been produced by plate movements and usually by the destruction of an ocean. Take one of the least probable-looking cases: the Ural mountains that divide Europe from Asia, running north and south through Russia. For the most part they lie far from any existing ocean. But now compare them, in that respect, with the Himalayas. The clear evidence of recent plate movements tells us that India has collided with Asia during the past fifty million years and made the Himalayas. The ocean that lay between India and Asia has vanished; instead, there is a chain of mountains lying deep inside an enlarged continent. Exactly the same kind of event can explain the Urals, with a collision between Europe and Asia. But the Urals were made around —225 MY, and the union of Europe and Asia at that time was apparently the last big step in the creation of Pangaea.

If you need an ocean to explain a mountain chain, and if an ocean is not already handy as it is in the Andes and the coast ranges of western North America, you must tear along the dotted line of the mountain chain to open a former ocean and delineate the pieces of continental material that collided to make the mountains. Some of the older mountain chains have been broken during the fragmentation of Pangaea. Alfred Wegener knew, for example, that the older mountains of Europe were matched to those of eastern North America and formed continuous ranges that split only when the Atlantic opened. But to explain why these mountains were there in the first place, it is necessary to invoke not one but two oceans that were long since destroyed in a collision of continents.

Before engaging in the strange mobile geography that made the Appalachians of the United States, and the Scottish Highlands and various other well-known features of European scenery, we should note the essential mechanisms, because it is from them and their consequences that the proof must come. In what follows, I draw on the ideas of John Dewey and others.

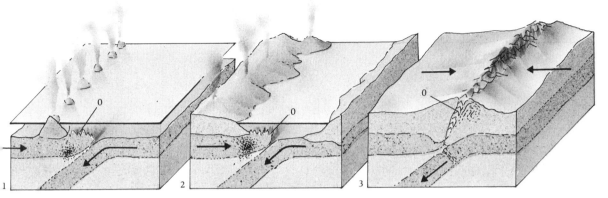

If an ocean trench lies beside a pre-existing continent, the oceanic plate usually dives to its destruction under the continental margin, as in western South America today. As the oceanic plate goes down, the continent scrapes some of the sediment from it and also tears some hard rock from the upper surface of the plate. The latter supplies the peculiar oceanic rocks of the kind described at Betts Cove at the beginning of the chapter. The characteristic collection of pillow-shaped lavas, sheets of basalt and heavy rock brought from the depths is called an 'ophiolite suite', but I shall refer to the rocks as ocean-floor markers.

A somewhat different geological environment prevails when the trench lies far from a continental mainland and one oceanic plate is driving under another. Then the result is an island arc, like Japan or the Aleutians. Volcanic islands emerge but their emissions are less 'continental' in character than those of the Andes. The land of the arc collects scrapings from the ocean floor, much as the leading edge of a continent does.

In either case, the sediments and the ocean-floor markers undergo deformation and heating, to an extent that depends on the relative speed of the downgoing plate. That plate also heats the continental rocks from below, causing volcanic eruptions and the formation of big blocks of granite.

Now imagine an oceanic plate going down under a continent or island arc, but carrying on its own back another continent. This approaching land cannot go 'down the drain' along with the oceanic plate; it is made of relatively light rocks and it is too buoyant. It therefore collides with the first continent, and the result is a crushing more severe than anything generated by the ocean floor alone. The mass of sediments caught up between the two continental edges is squeezed upwards and tends to spill over on to the land of the downgoing plate. By the collision, the continents

are welded into a bigger unit. Eventually the trench underneath them gives up trying to swallow them and the Earth's shell breaks at another point.

If some of the ocean-floor markers reach the surface of the resulting mountains, or if they are exposed by later erosion, these markers show the joint-line between the two continents and represent the last visible remains of the ocean that brought them together. Such is the case in the lower slopes of the Matterhorn (see p. 34). Similar markers occur in the Himalayas and other young mountain chains. More significantly, they are found in much older mountain chains, including the Urals and the Atlantic chains mentioned earlier, thus indicating that the same kinds of processes were at work before the time of Pangaea.

Even for mountain chains having no identified ocean-floor markers it is usually reasonable to assume that they, too, formed in much the same ways. But there is a choice of mechanisms in plate tectonics for making mountains and it is not always obvious in what circumstances, or on which of the moving continents, a particular mountain chain was built.

In a few places, including the young southern Rockies of the USA, mountain-building seems to have occurred far from any ocean. Apparently continents can divide into two plates which then, instead of parting, move towards one another and buckle their margins. But the Rockies, together with the young inner chains of the Andes of South America, are products of complex events going on while the American continents overrode the floor to the west – events that the new plate theory has not yet disentangled adequately.

Finally, there are paradoxical mountains created during rifting rather than by the convergence of plates. In the East African rift zones the mountains rose on huge 'bubbles', or domes, formed by the heating from below that preceded the rifting.

The founding of Pangaea

Geological reasoning and information from rock magnetism give a rough synopsis of events in the period from 600 to 200 million years ago; rough, because only now are the first serious attempts being made to reconstruct this complicated era of the Earth's history. The earth scientists are in much the same position as the first globe makers were, who tried to map the world from the scanty information brought back by seafarers. There will be argument about the details for years to come, even among those who are agreed that continental fragments were adrift during this period.

From well before —600 MY, major continents were apparently in continuous motion relative to one another. There followed three important episodes of mountain-building associated with continental collisions: one at about —420 MY between North America and Europe, another at —300 to —250 MY between Africa and the joint Euro-America, and the third at about —225 MY between Asia and Europe. These were the main events in the founding of Pangaea. In addition, the continents were colliding with island arcs while minor continental fragments (of east Asia in particular) moved independently. For the sake of clarity, it is better at this stage to disregard these lesser changes; otherwise the maps and motions become confusing.

The states of the world shown on the accompanying maps are those deduced jointly by James Briden, a rock magnetist at Leeds University, and Alan Smith of Cambridge University. At —500 MY, the map of the world was odder even than at the time of Pangaea. Most of the continents appear to have been massed together in an enlarged Gondwanaland. It consisted of South America, Africa, India, Antarctica and Australia, and also the main body of Asia, which may have been about to break away at that time. This whole assembly was upside-down, with North Africa near the South Pole.

The amassing of that supercontinent may have had important consequences for life on Earth, as we shall see later. At any rate, during the era represented on the maps, complex plants and animals were evolving rapidly and life came ashore for the first time.

The territories of the continents were different from those of today. Here, too, it probably pays to overlook those differences, except in one interesting respect: a substantial part of present-day Europe was attached to North Africa.

The rest of Europe, together with North America, was detached from the supercontinent. Here the geological and magnetic evidence are at odds. The latter allows for very little relative motion between Europe and North America. Yet there was fairly obviously a collision between them at about —420 MY, as already mentioned, which made the mountains of Norway, east Greenland and Scotland. The former ocean that separated them may have been no wider than the Mediterranean is today.

Meanwhile, Gondwanaland was putting itself right way up, probably by running across the south polar region to the other side of the world. Asia travelled by itself. The situation at —400 MY shows Europe and North America firmly welded together but still well out of place, while Asia and Gondwanaland were already close to their eventual positions in Pangaea. By —320 MY, Euro-America had moved into a position where two further collisions became possible.

The first, with Africa, made mountains that run from east to west across the middle of Europe and through the eastern United States. The second, with Asia, was responsible for the Ural mountains that now run from north to south across Russia. Pangaea was thus completed but it was a fleeting assembly: its break-up began about 40 million years after its completion.

Continental movements before the supercontinent of Pangaea came into being, as plotted tentatively by J. C. Briden and A. G. Smith. In the first set of globes (500 million years ago) the gap labelled 'c' between North America and Europe was a temporary and narrow one, but its closure made the mountains of Norway and Scotland. Note also the attachment of southern Europe to Africa. The last set of globes (320 million years ago) shows the state of the world just before the collisions between Europe, Africa and Asia that completed Pangaea. For further description, see the text.

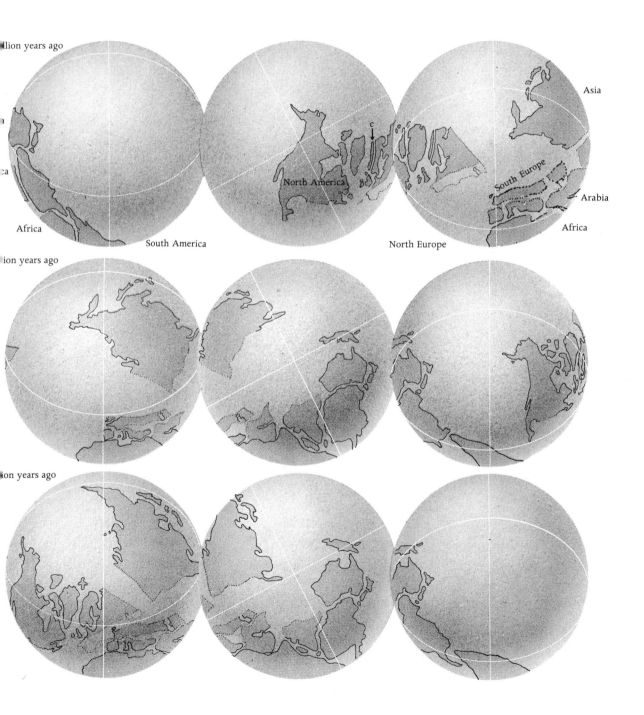

The Mountains of Norway were made in a collision between North America and Europe, about 420 million years ago.

Precambrian rhythms

In mountain chains older than about 600 million years, no convincing ocean-floor markers have yet been discovered. Conceivably the composition of the ocean floor itself was different before that time and different materials will have to be sought as ocean-floor markers. Yet, going back even more than 2000 million years, mountain belts appear to be essentially the same structures as young mountain belts. Although they are much eroded, they show the same kinds of ordinary rock types, in formations running in long lines and with evidence of overthrusting similar to that occurring in the Alps.

These old rocks and mountain chains make up much of eastern Canada, the northern Baltic lands of Europe, and large parts of the USSR, Africa and Australia. In many other regions they lie covered by younger rocks. Being barren of fossils except of the most primitive forms of life, they were all lumped together as 'Precambrian' rocks by the nineteenth-century geologists, who did not realise that they spanned a far greater part of the Earth's history than all the intricately divided geological periods since the start of the 'Cambrian' at −600 MY. The timeworn Precambrian regions have been illuminated by the development of atomic techniques for discovering the ages of formation of the rocks (see Chapter 4). These dating techniques have enabled earth scientists to map 'provinces' of old rocks and mountains of different ages.

The oldest rocks of all are usually very much deformed and reworked, suggesting that there were no sizeable, stable units comparable with present-day continents until about −2700 MY. If, thereafter, continents were moving around, colliding to make new mountains and breaking up again, for 2100 million years, our present names and outlines for the continents are largely meaningless. Instead we must visualise frag-

A very old fault-line in the Canadian Shield. This cliff, 900 feet high, is the boundary between two 'provinces' of old rock. Those at the left of it are about 1900 million years old; those on the right, 2800 million years. The rocks represent nameless continents that wandered about the Earth for unimaginably long periods, when life in the sea was still rudimentary.

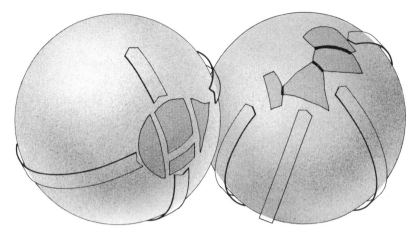

John Sutton's concept of a succession of pangaeas in the early phases of the Earth's history. The pieces of a disintegrating supercontinent will tend to reassemble, but in a different configuration.

ments of the present continents, such as the Congo region of Africa and the Superior province of Canada, moving independently or in bizarre unions with other old provinces that are now on the other side of the world. Because of the tendency of continents to split along old seams, the provinces may have retained their identity for long periods.

The photograph shows a conspicuous boundary in Canada, where rocks differing in age by nearly 1000 million years lie side by side. Unfortunately, information about these provinces is often buried under newer mountains or the thick ice sheets of Antarctica and Greenland.

To reconstruct their journeys will be difficult. Investigators of rock magnetism are only just beginning in earnest to attempt the Precambrian and many years of work will be needed, especially as the permanence of the continents is in question and each province needs to be considered separately.

Meanwhile, the ages of provinces gives us pieces of a puzzle that could be rearranged in many different, equally plausible, equally unreliable ways. But a striking feature of the Precambrian ages serves as a guide to thought. Creation of new continental masses did not occur uniformly throughout this unimaginably long period. Instead, mountain-building was episodic, even rhythmic.

The ages of the known provinces of the Precambrian, excepting the very oldest, tend to cluster in three intervals beginning at about —2800, —1900 and —1100 MY. In between these periods, the rate of formation of durable continental material was much reduced. Indeed, in many places nothing seems to have happened during the lulls, and rocks of these main mountain-building periods lie side by side. If we look at more recent mountain-building with an out-of-focus lens, so that we see only broad, imprecise strokes equi-

valent to what we make out in the Precambrian, it too is episodic.

John Sutton of Imperial College, London, is led to envisage a cyclic process of continental movements, making and breaking a succession of pangaeas. Consider the present situation on Earth: the Pangaea of —200 MY has broken up and the pieces are scattering. But, because the planet is a sphere, the present continents will tend to meet one another coming around the other way, on the far side of the Earth: if the Atlantic and Indian oceans continue to grow, the continents will merge again into another supercontinent over the ruins of the Pacific – in, say, 200 million years time. The making of the Alps and the Himalayas are early steps in this new merger. One effect will be to turn everything inside out. Regions such as the western USA and China, which were on the outside of 'Pangaea —200', will be deep in the interior of 'Pangaea +200', while former heartlands of the old supercontinent, the eastern USA and West Africa for example, will be coastal regions of the next supercontinent.

Just such a process explains, for Sutton, the episodic mountain building of the Precambrian. He imagines that it occurred most vigorously at the times when continents were reuniting in a supercontinent. The lulls would have been periods of stability or break-up of supercontinents; during them, some mountains continued to form on the leading edges of the continents, as in the Andes today.

Plate-made world

The French philosopher Pascal remarked of Cleopatra's nose: 'had it been shorter, the whole history of the world would have been different.' We can go further. Had the rift zones of Africa not heaved up mountains there would have been no mighty Nile, no Pharaohs,

Drilling for oil in Alaska. The oil was formed 200 to 300 million years ago, when Alaska lay more to to the south.

and Cleopatra would have been a desert nomad. If India had voyaged elsewhere, China would not have been hemmed in by mountains and its modes of civilisation and political theory would not have been so distinctive. If the Atlantic had opened along a slightly different rift, Ireland could have remained attached to Greenland, with consequences for the history of Britain that I leave as an exercise for the reader. Americans may like to speculate about the similar missed opportunities of leaving Boston in the Sahara and Los Angeles in Mexico.

If the great diversity of human ways of life is counted as an asset, we are fortunate in our present geography. The unusual number of continents and islands created by the last break-up has given us an immense variety of niches. Pangaea would have been a dictator's delight, a perfect setting for the pan-continental empire.

In our trigger-happy world, political geography is uppermost in most people's minds. It is difficult to avoid finding metaphor in the present plate movements, which are widening the Atlantic, taking Australia rapidly closer to China and promising to create a Euro-African condominium. But such trains of thought must always reach the same conclusion which spoils the joke: that plate movements are extremely slow compared with the span of a human life, or of all human life till now. The Atlantic has widened by only a few yards since Columbus crossed it and Australia may take 100 million years to reach China.

Nevertheless, ongoing plate movements have occasional profound effects on human affairs, when the toll of death by earthquake or volcano becomes politically significant, as with the explosion of Thera-Atlantis which altered the balance of power in the Aegean region thirty-four centuries ago. The repeated disruption of Middle Eastern irrigation systems by earthquakes cannot have aided prosperity in that region. At least one soothsayer of ancient China ascribed the fall of the dynasties to earthquakes causing interruption of the rivers on which agriculture depended. Other, less catastrophic, processes have had subtle effects which historians might well investigate, including the changes in level of the Mediterranean shores during and since the ascendancy of Greece and Rome.

The chief influence of geology on human affairs is economic, especially in connection with minerals, with fossil fuels like coal and oil, and with farming soil. A brief word about soil and the influence of the rocks from which it is derived: climate and vegetation can make the same good or poor kinds of soil from different kinds of rock, but volcanic rock weathers more rapidly than granite does, giving a better supply of major plant nutrients. Agricultural scientists have discovered, in recent decades, the importance of 'trace elements' – materials such as zinc, cobalt and molybdenum which must be present in small amounts in the soil if plants and animals are to be neither deprived nor poisoned. Limestones, for example, make soils that tend to be peculiar in trace elements. Here, as in the bearing of trace elements on geographical patterns of human disease, knowledge is at present sketchy.

Only now are geologists beginning to work out how the newly discovered plate actions help to account for the location of valuable minerals. The simple fact of continental drift answers the question of how there comes to be oil in the Arctic and coal in Antarctica, materials derived from plant growth in sunnier climates: Alaska and Antarctica have spent periods of their existence in the tropics. An important factor in economic history has been the concentration in the northern hemisphere of the most accessible supplies of coal and oil. During the period of formation of most

Terraces of carbonate rock in western Turkey, made by hot springs. Heating due to activity at plate boundaries causes important chemical changes in the Earth's surface, and helps to make minerals that are useful to man.

coal and oil the direction of continental movement has been generally northward, and the coal miners of Pennsylvania, South Wales and the Ruhr harvest the remains of former tropical forests.

Basalt into ore

Every important deposit of materials of interest to man represents some kind of geological accident. Nevertheless the production of most minerals seems to be a consequence of plate movements. The formation of diamonds is somewhat exceptional. They crystallise at high temperatures and pressures deep underground and are propelled upwards by an explosion that bores a 'diamond pipe' towards the surface. The chief sources of diamonds are in old Precambrian rocks, as in South Africa and Russia, and there is some argument among geologists about whether the Earth has given up making diamonds, or whether younger diamond pipes have not yet been exposed by erosion.

The same controversy surrounds gold ores, which again are mined mainly in Precambrian deposits; some geologists say that gold found in younger rocks is merely Precambrian gold reworked in later events, while others declare that the Earth is still quite capable of making gold ore. The compromise is to say that there may have been a gradual thickening and some chemical change in the plates and the material they carry, during the Earth's long history. Apart from any other consideration, the chemical make-up of the atmosphere has changed, while the loss of material to the enlarging continents must have altered the internal composition of the planet.

Gemstones cast a sidelight on some of the processes involved in the production of metal ores. Where the Earth's shell splits during the opening of an ocean, allowing material from the deep interior of the Earth

to come to the surface, one of the chief ingredients of that material is olivine; the island of St John in the Red Sea, where a new ocean is opening, is the source of the finest olivine of gem quality, known as peridot.

Where an oceanic plate is consumed under the margin of a continent it creates, at some levels, a high pressure but at a comparatively low temperature. In these conditions various transformations occur, one of the most telling being the conversion of a common material, albite, into jadeite, including the preferred form of jade. The Chinese obtained their jade from Mogok, now in the heart of Burma but formerly on a continental margin or island arc. The Aztecs of Mexico also worked in jade; jadeite occurs extensively down the west coast, or leading edge, of the Americas. Higher temperatures created the famous emeralds of Colombia, which drove the Spanish conquistadors almost insane with avarice. Continental collisions form garnets from common materials by high pressure and temperature. In due course, attempts will no doubt be made to map past and present plate boundaries from the occurrences of different kinds of precious stones.

Many metal ores, too, are formed where plates are growing or being consumed. As one of the world's youngest and narrowest oceans, the Red Sea has curious pools on its bottom that are filled with very hot, very salty water. At the bottoms of these pools, rich deposits of metals have accumulated, especially of iron, manganese, zinc and copper; the deposits may be up to 300 feet thick. Investigations during the past few years, especially by American oceanographers from Woods Hole, Massachusetts, have explained the origin of the brines in hot springs on the ocean bed. These springs bring water that has percolated through the young rocks of the Red Sea crust, dissolving the metals as it went. In the Red Sea, the process is greatly aided by a

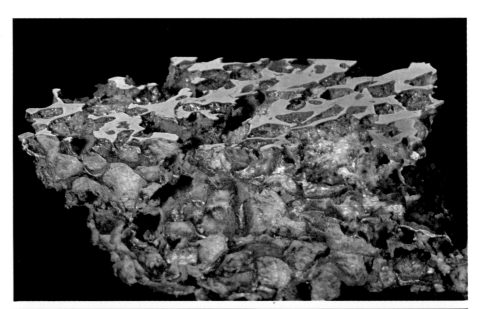

slope of the ocean floor, but the water may have left most of its charge of metals in veins under the ocean floor.

In this leaching of new volcanic rock by water, under the Red Sea, we may be seeing a process that has often operated in the past, making many of the Earth's best ores. Recently, earth scientists have begun to examine deep-lying regions of the continental crust with some precision, by echo-sounding with man-made waves. In Alberta, Canada, this method of exploring the crust shows what is apparently a rift about 30 miles wide in old rocks that are now buried under a great thickness of younger rocks. An important deposit of lead-zinc lies in the line of the rift, so it is tempting to imagine something like the Red Sea process operating in Alberta 1500 million years ago – and in many other hidden rifts in other parts of the world.

The birth of an ocean may be the most fruitful season for the making of ores. Materials formed at that time will remain near the edges of the continents, as the ocean grows from the middle outwards. Salt domes at the sides of young oceans may be traps for oil. But chemical activity continues at the mid-ocean ridge and on its flanks.

The newly formed ocean floor is coated with large amounts of metal compounds, notably sulphides, derived from the hot upwelling material, primarily basalt, at the plate boundary. Quite often ores of copper, lead and so on form neat layers; the layers are thin but the area of the ocean is eventually immense. There is also interaction with the ocean itself: organic sediments that rain down on the floor are relatively rich in certain metals, such as uranium and molybdenum. During the life of the ocean floor, its uppermost layers lose some constituents to the water and gain others from the water. These exchanges help to keep the chemical composition of the ocean water remarkably

constant. Among their products are the peculiar nodules of manganese that litter large areas of the ocean floor and are now exciting the interest of would-be ocean miners.

When, in due course, the oceanic plate dives to destruction in an ocean trench, it takes with it the greater part of its charge of metals. But some of the upper layers of the plate are scraped off, and heaped up on land, as in the copper mines of Newfoundland 500 million years ago or – to give a topical example – on Bougainville Island, near New Guinea, today. A copper mine containing 900 million tons of ore, and beginning production in 1972, stands beside an active ocean trench.

The heat generated in the destruction of oceans and continental collisions has important chemical effects. Many major deposits of copper bear signs of having been caught up in the roots of former volcanoes. The tin of south-west England, which drew traders from far afield in classical times, comes from the granite masses formed by heat during the first collision between Africa and Europe. The tin of Malaysia and Indonesia is associated with granite made in plate movements so far unspecified, 150 million years ago.

The formation of ores is not as automatic as this sketch may suggest. Each important ore body requires some special local or regional event to produce it – depending perhaps on water trickling through volcanic rocks, perhaps on some climatic factor, perhaps even on biological action. And these events may occur long after the original addition of oceanic rocks and sediments to the continental fabric. Thus it is not easy to translate present activities at plate boundaries into a formula for the location of ores, except in the most general sense. Plate boundaries supply the heat and important chemical ingredients, so metal ores tend to be located at past or present plate boundaries. But then every scrap of land has probably been close to a plate boundary at some time or another.

In one of the clearest cases, we can say that the vast metallic wealth of the west of North America and South America – silver, copper, zinc, lead, tin, iron, manganese and so on – originates in the prolonged overriding of the plates of the Pacific Ocean, either by those continents themselves or by island arcs with which they later collided. In Wales, mineral prospectors are showing renewed interest in Snowdonia, which was formerly an Andes-like volcanic region on the edge of Europe.

Prospectors may find the knowledge of recent plate movements helpful in another way. Proof of continental splitting leads at once to the idea that areas rich in minerals may have been cut through by the rifts. The coastal area of West Africa, where Brazil formerly lay, carries valuable minerals, including ores of zinc, tin and gold, and also diamonds. So there may be a case for looking more thoroughly at the minerals under the Amazon jungle, to see whether similar deposits occur. As a small token of things to come this reasoning has already been effective in the discovery of salt in Brazil, a commodity in short supply in that country; it followed the discovery of salt domes and oil in Gabon, West Africa.

The catch is, of course, that the effect of these aids to the discovery of new mineral deposits is to help us make a thorough job of finding and exhausting the ores of the Earth. The present exploitation is spendthrift, to put it mildly. The easily accessible deposits of rich ores, accumulated in many millions of years of Earth history, are disappearing in the course of a few centuries. This is a theme to return to later, in considering man's present and future relationship to his planet.

Most of this book takes a global view of past and present events in the shaping of the lands we live in.

That is the nature of the new geology. The Alps were made because the Atlantic opened; the earthquakes of Japan are due to the motion of a plate that grows from a ridge off Central America, and so on – plate activities span the continents and oceans. Nevertheless, this global view, with endless references to faraway places, can make the story seem abstract and may allow us to overlook the fact that the ground under our own feet was made by these selfsame processes of plate tectonics. So it is worth devoting a few pages to scenery familiar to the reader.

The new story of America

In some respects the USA and Canada are ideal for demonstrating the workings of the new geology, in explaining adequately, for the first time, the grandeur and diversity of the face of the continent. The opening of new oceans; the destruction of old oceans at ocean trenches causing volcanic island arcs and collisions of continents; the long-lasting sidling action of the transform fault in California – all the processes of plate tectonics have combined to make North America the way it is. Difficulty arises only because these great events have occurred repeatedly in relatively narrow belts, so that the evidence is somewhat confused. At this very early stage of the game of reinterpretation it is not surprising that there are conflicting opinions. This first attempt at a synthesis of theories draws on the work of a number of earth scientists.

For most of the Earth's history, North America did not exist in any recognisable form. The oldest known rocks are in Greenland (up to 3900 million years old) and in Minnesota (up to 3350). In the United States, this region of old rocks extends into Michigan on the surface and it continues westwards underground towards Idaho. Northern Europe was probably part of

Old, worn-down rocks of Northern Labrador. The Canadian shield, extending into the north-central states of the USA, is the ancient core of North America.

the same old continent and by —1000 MY North America was extended or remodelled in a broad belt of territory stretching from the Baltic Sea in Europe, via Labrador and Quebec, right across the central USA. But that hazy era of continent-building is, we have already confessed, beyond precise reconstruction at present.

Let us take up the story at around —500 MY. North America and Europe, looking very different from their present configurations, lay close together in the tropics. North America was rotated so that what is now the east coast ran roughly parallel to the Equator. This was in a period of break-up of a supercontinent and a narrow ocean, perhaps no wider than the Mediterranean, separated North America from its European partner. We must visualise bare rock, for life had not yet ventured out of the water.

The shaping of the continent proceeded on all its coasts, but it is convenient to begin with the east coast, at the south-western end of that narrow ocean between North America and Europe. Around —470 MY, an arc of volcanic islands was heaped on to the shore of Quebec along the line of the St Lawrence river, supplying the mountains of Newfoundland, New Brunswick, and Maine.

That was only a preliminary to a greater event, in which North America and Europe collided violently, at about —420 MY. The collision produced a chain of mountains like the present-day Alps stretching along eastern Greenland to Newfoundland, and incorporating Norway, the Scottish Highlands and the north of Ireland.

Europe and North America were thus welded together again. For more than a megacentury (100 million years) they were wheeling and drifting slowly in the tropics, undisturbed by other major continents, although at about —380 MY an east-coast collision with another island arc formed the mountains of Vermont and made other additions to the Northern Appalachians. The sea-level was high; depressed parts of the dual continent were flooded, allowing sandstones, shales and limestones to accumulate. On land, life was beginning to prosper and great jungles of trees and ferns cloaked the previously barren rocks.

Then Africa moved in and rammed both North America and Europe. Before —250 MY, the impact threw up the main ranges of the Appalachians. The eastern coastal states from Georgia to Maine were, in the process, joined to North America for the first time – a gift from Senegal in West Africa. Birmingham, Alabama, stands at one end of a chain of mountains whose other end is in eastern Europe. This great event required the destruction of an ocean between Africa and North America, and the disappearing ocean floor created volcanic mountains on the African side – for example, those of central North Carolina. The same movement brought South America, Africa's long-faithful consort, alongside the Caribbean region.

Thus was North America incorporated on the edge of the supercontinent of Pangaea, which, as it formed, began drifting slowly northwards. Its eventual break-up fashioned the familiar eastern outline of the United States. The Atlantic opened between North America and Africa around —190 MY. At about the same time, Nicaragua and Yucatan parted company with Louisiana and Texas, before taking up their positions in central America. The unzipping of North America from Europe was not completed until more than 100 million years later. The early rupture of Pangaea is marked by the outpouring of molten rock that made the Hudson Palisades of New York and New Jersey, and by evidence of volcanic activity in the White Mountains region of New Hampshire.

The western part of the old core of North America

lay on the outside of the Pangaean supercontinent and so remained exposed to maritime events throughout the past 500 million years. What is now the eastern end of Russia was a part of North America, attached to Alaska. But Alaska itself was closed towards the northern islands of Canada, so that the west coast of the continent was more nearly straight than it is now.

From where the Rockies rise quite abruptly at the western edge of the Great Plains, all the way to the Pacific coast, there is a succession of ranges running at least roughly parallel to the west coast of North America. The states of Oregon and Washington, and British Columbia and the Yukon in Canada, are evidently additions to the continent. Looking back, we must envisage a continental margin in Nevada, Idaho and Alberta, with oceanic plates being destroyed along it, and the continent being enlarged and crumpled in several episodes. Sometimes the floor of the adjacent ocean would be diving to destruction under the continental margin, creating volcanoes and compressing the continent, as in the Andes today. At other times, the continent itself would be dipping on the descending plate and heaping on its leading edge oceanic rocks and occasional island arcs.

Early mountains, eroded away and eclipsed by subsequent events, are often identified only by the sharp, analytical eyes of the field geologist. That certainly applies to the so-called Antler event of about −350 MY, the existence of which went almost unrecognised until 1942. An encounter with an ocean trench made a range of mountains running from central Nevada through Idaho into British Columbia. A second such episode occurred around −220 MY, with the formation of the Cassiar mountains of the Yukon and British Columbia, an event which also added to the continent a mass of oceanic rocks represented in the Canyon Mountains of Oregon. A third collision with an ocean trench took place at about −100 MY; the region of the Sierra Nevada was folded, and a deep-lying mass of granite was formed which was, much later (−2 MY), heaved up to form the present backbone of the Sierra Nevada.

Meanwhile, just before −100 MY, another important event had occurred. Alaska and its 'Russian' appendage collided with the main continent of Asia, making the Verkhoyansk mountains of Eastern Siberia. There followed a wrenching action which twisted Alaska and its mountains anti-clockwise and made them jut towards the west.

Intermittently, throughout the history of the west, great volcanic activity occurred as the continent overrode the adjoining ocean floor. The Coast Ranges, from the Yukon to California, represent at least in part oceanic material which was heaped on to the continent by a plate descending under the continent at an ocean trench, at about −100 MY. There is much ocean-floor rock in California. That underthrusting by the ocean continues in Oregon and British Columbia, marked by the continent's active volcanoes. Further south, the recently dominant action has been that of transform faulting along the San Andreas axis. As described earlier (page 60) associated tugging is thought to have made the 'Basin and Range' province of the southwestern USA, in which great blocks of ground have rifted and subsided as the land has been stretched. The San Andreas transform fault is still active, so the shaping of North America is unfinished.

The Rockies, North America's greatest chain of mountains, remain something of an enigma. The more southerly young Rockies of Wyoming, Colorado and New Mexico were formed in the abolition of a seaway dividing the continent, during the past 70 million years. The problem is to explain why this relatively young mountain chain occurs inland, 800 miles from the battleground of the western seaboard. It may be

Yosemite National Park. In California, a long lasting tussle between plates has built mountains, and the recent ice ages have eroded them, making glacial valleys.

that the 'soft' continental plate broke at the western edge of the Great Plains and that the eastern part began to thrust itself under the western part.

Nor is there any very obvious reason why the Colorado Plateau (of Colorado, Utah, Arizona and New Mexico) should have bobbed more than a mile into the air during the past few million years. For the time being, therefore, America's most famous geological feature, the Grand Canyon, carved out by the Colorado River as that plateau slowly rose, eludes an easy explanation by plate tectonics. One guess is that the plateau is an after-effect of the swallowing of a piece of the 'mid'-ocean ridge of the Pacific, under the American continental margin.

Throughout this long story of North America, wind and water were wearing away mountains and hills while rivers carrying silt were building up the land elsewhere. The weather has intervened most dramatically since —2 MY, in the present series of ice ages. At its greatest extent, the arctic ice sheet covered the continent as far south as Seattle in the west and Indianapolis in the east, supplemented by miniature icecaps on the Rockies and other mountains. Ice shaped many valleys of the western USA, of which the most celebrated is the Yosemite valley in the Sierra Nevada of California; its waterfalls mark the places where the main glacier was fed by tributaries.

The Great Lakes have been made, obliterated and made again by successive ice sheets; the lakes formed during the melting of the ice. In North Dakota and Manitoba, wheat grows on what was, a few thousand years ago, the bed of another huge lake; Lake Winnipeg is a much-shrunken relic of it. Canada is still rising, following its last relief from the load of ice, and it will tip the water of Lake Michigan into the Mississippi. The ice will probably return many times during the next few million years.

The restless Earth will have other events in store for North America. If present motions continue, the Pacific plate will carry away south-west California to be an island off Alaska. Something will have to give in Alaska or Siberia, because North America, after breaking from Eurasia in the east, is still bearing down upon it in the west. For as long as the Atlantic continues to grow, the eastern side of the continent will be geologically tranquil, but sooner or later – millions of years from now – there will be a new ocean trench in the east, with a 50 per cent chance of volcanoes in Charleston and New York City.

*The Greenland site of the oldest known rocks on Earth – about
3800 million years of age – and their discoverer, Vic McGregor.*

Chapter 4   The Rocky Motor

Eventually the history of the Earth and the machinery that now drives the plates will both have to be explained in terms of the materials that first came together to make the planet, and their subsequent behaviour. Telescopes, moonships and powerful presses are all yielding relevant knowledge about the origin and evolution of the Earth.

A young New Zealander working for the Greenland Geological Survey discovered the oldest rocks on Earth. In 1966–7, Vic McGregor deduced that certain rocks in Greenland were exceptionally old but several years elapsed before his conclusions were confirmed by atomic dating methods. By 1970 he had made up his mind to build a house in a remote village on the coast of Greenland, to learn the language from the Greenlander children themselves and then to set up a little school. But McGregor did not abandon his research entirely, and word came from Canada and from England that McGregor's rocks had at last been dated. They were 3700 to 3900 million years old – an age greater than any determined on this planet before.

These venerable rocks occur in western Greenland at the mouth of a fjord, Ameralik, not far from the capital, Godthaab. McGregor was led to recognise them as being very old by his reasoning as a skilled field geologist. His chief clue, when he examined the Ameralik rocks, came in bodies of dark rock that had invaded light-coloured rock. He deduced that there had been at least two major phases of intrusion. Some of the intruding material was itself very old and the light-coloured rock which suffered the intrusion was therefore even more ancient. It did not, though, appear to be different from many other rocks; nor did it fit most geologists' conception of what very old rocks should look like.

At the time of McGregor's discovery, the oldest known rocks were in South Africa, in a region of the Transvaal known as the Barberton, and were dated at −3400 to −3500 MY. Researchers at the Universities of Alberta and Oxford, using two quite different dating techniques, found the Ameralik rock to be substantially older.

The techniques for dating rock have only recently become adequate to cope with material as old and con-fused as the Ameralik samples. Every time a rock is affected by some upheaval, changes are liable to occur in it which alter its apparent age; the problem is, as it were, to 'see through' these intervening events to find the real age of the rock. Dating is done by atomic methods which give a precise figure, but the precision can be misleading, and different techniques can give very different answers. A great deal of geological reasoning is required to select the right technique and to decide what the figures really mean.

The principle of atomic dating is simple enough. It relies on the fact that almost all materials contain a quantity of radioactive atoms. Each class of radioactive atoms decays at an exactly predictable rate and in the process it makes, either immediately or by a series of steps, a stable 'daughter' atom which will not decay. As time passes in an undisturbed rock, the amount of the radioactive material diminishes and the amount of daughter material increases. The relative amounts of atoms of different elements, or of different weights within the same element, show how much time has elapsed since the radioactive 'clock' was set running. Ernest Rutherford, who discovered the law of radioactive decay in 1902, saw that it could be applied in dating minerals. Those who took up the idea with great effect included B. B. Boltwood and Alfred Nier in the USA and Arthur Holmes in Britain. Since then, techniques have advanced beyond recognition and many other sciences exploit radioactive atoms for measuring time, including the carbon-14 of the archaeologists.

The radioactive elements most often used as clocks in geology are uranium (U-238 and U-235 which decay into lead), rubidium (Rb-87 which changes to strontium) and potassium (K-40 which yields argon gas).

At Oxford, the rock-dating team in the department of geology chose the uranium clock for measuring the age of McGregor's samples from Ameralik, but they

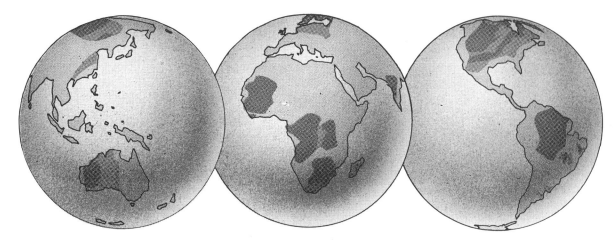

used it indirectly. With the help of a machine called a mass spectrometer, they compared the amounts of lead atoms of different weights. The decay of uranium produces relatively heavy atoms of lead so, in the course of the Earth's history, the atomic composition of lead changes. (Incidentally, a serious practical difficulty in exact work of this kind is avoiding contamination by the 'modern' lead injected into the air of Oxford by car exhausts.) Afterwards, the team used the rubidium clock and deduced an age that was, if anything, greater. They measured the accumulation of strontium atoms attributable to the decay of rubidium. The raw results gave ages close to 4000 million years for McGregor's samples, but there is a margin of uncertainty about the 'rate' of the rubidium clock.

With atomic methods of dating, the geologists can now put figures to periods and events that were formerly known only as 'Jurassic' or 'Albian' or by other colourful but uninstructive names. The benefit of dating techniques has been especially great in the study of rocks formed before —600 MY in which there are very few fossils for helping the geologists to find contrasts between rocks of different ages. Conclusions from dating have been incorporated in many of the accounts of continental movements and collisions that have figured in earlier chapters.

In Greenland, Vic McGregor is continuing his investigations of the very old Ameralik rocks. For earth scientists they are more than mere entries for the record book. They should help to disclose the state of the material from which the Earth's continents first formed and that, in turn, will be an invaluable guide to the workings of the rocky motor of the Earth's interior. Any satisfactory explanation of the features and movements we see today must tell a sequential story from the very creation of the Earth, in terms of the raw materials which went into it.

The birth of the planets

Astronomers have never watched a planet being formed but in 1936 they did, by chance, spot a new star bursting into light. More recently, very young stars have been investigated in various parts of the sky. They flash irregularly, a sign of upheavals that probably eject great quantities of matter. Dust and gas surround them and analysis of the star light reveals the presence of complex materials of the kind from which planets may eventually form. These stars give an impression of what our own star, the Sun, must have been like in its infancy. They reinforce a long-held belief of astronomers that the Earth and other planets were made from a disk of matter swirling around the Sun soon after its own birth.

The raw material of the universe is hydrogen, the lightest element, and the cloud of gas from which the Sun originally condensed was nearly all hydrogen. Had it been pure hydrogen, we should not be here. The gas in our arm of the Milky Way was already contaminated with heavier elements made in other stars that formed, grew old and exploded before the Sun came into being. As stars burn, and especially as they go into dramatic throes when they run out of fuel, a complex sequence of nuclear reactions makes all other chemical elements known to us, out of the primeval hydrogen.

The making of the planets was in part a matter of distillation. Under the combined effects of gravity and heat, the different bodies of the solar system retained the chemical elements in different proportions. According to calculations of Fred Hoyle of Cambridge University, about one hundredth of the mass of the Sun was needed to make the planets. Most of that was thrown away, back into interstellar space, leaving some of the planets with concentrations of elements very different from the Sun. Magnetic forces from the Sun helped to spin off the material, but as Hoyle comments,

The oldest parts of the Earth's continents (left). The darker shading shows rocks older than 1700 million years (not necessarily on the surface) and the lighter shading, rocks of 800 to 1700 million years. The information for the USSR and China is uncertain. (Adapted from a map of P. M. Hurley and J. R. Rand.)

The surface of Venus (below), permanently shrouded in clouds, is unveiled by radar from Earth. The part of the surface revealed so far, and here positioned on the globe of Venus, suggest that continents and dry 'ocean' basins exist on Venus and that plate motions occur there. (California Institute of Technology.)

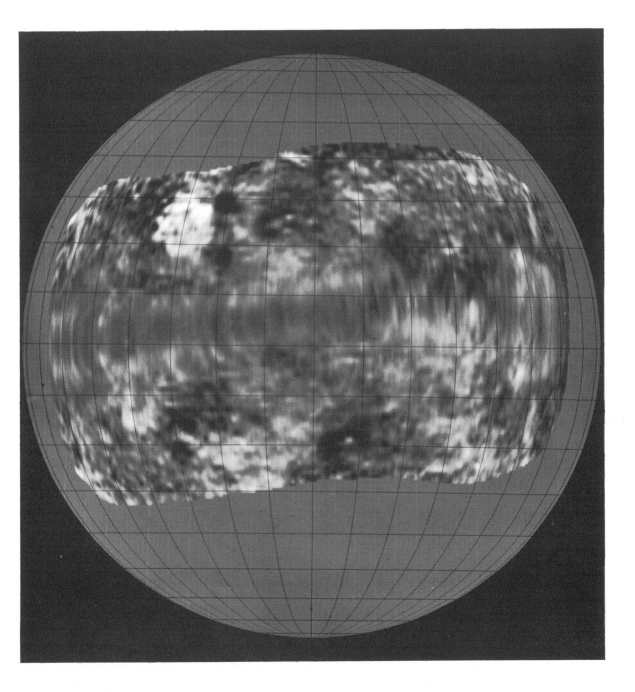

David Scott, commander of Apollo 15, *taking photographs of the lunar surface. In the distance are the lunar Apennines, which rise here to 15,000 feet. They look smaller and nearer because there are few clues to distance on the Moon.*

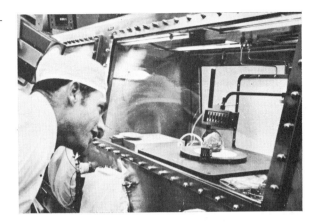

'One feels that if the dissipation had been only marginally more effective, all the planetary material would have been lost. The origin of the planets appears to be a matter of chance.'

The giant planets, Jupiter and Saturn, share the Sun's bias towards hydrogen – indeed, though too small to ignite, they are like miniature stars. The innermost planets, Mercury, Venus, Earth and Mars, formed in a region where, out of the vapours surrounding the Sun, key materials had condensed – particularly iron, magnesium oxide and silica, the prime constituents of the Earth, and possibly sulphur too. The material in the disk grew into planets in snowball fashion, with established boulders scooping up other material by chance encounters until they became massive enough to drag in copious supplies by gravity. The meteorites, lumps of rock and metal that still occasionally fall to the Earth from outer space, are a vestige of this process.

Of all the resulting planets, only Earth is comfortable to live on, at least for creatures like ourselves. Mercury and Venus are far too hot and the outlying planets are far too cold. The surface of Jupiter may be quite warm, but it is chemically unsuitable. The chief drawback of Mars is the lack of water, oxygen and nitrogen in its thin atmosphere, which consists almost entirely of carbon dioxide. But until sufficient unmanned spacecraft equipped with 'bug detectors' have landed on the planet, we cannot rule out the possibility of some kind of life existing on Mars.

Venus is the planet closest to the Earth in size as well as by distance. The solid parts of the planets are so much alike, in measurable size and density, that plate movements may occur on Venus as on Earth, although the oceanic basins must be empty of water. Thick clouds obscure the planet's surface, but radar from the Earth penetrates them, and gives the first glimpses of surface features that could be continents and mountains.

Geology on the Moon

When the astronauts David Scott and Jim Irwin came back from the Moon they did not have to go into quarantine as their predecessors had done. By then the Moon was declared free of viruses or spores that might infect the Earth. But the boxes of lunar rock they brought back with them in *Apollo 15* were still received and handled with gingerly precaution, lest they be contaminated by the Earth.

At the Lunar Receiving Laboratory at Houston, the main batch of samples went into an atmosphere of sterile nitrogen. From the outside, using 'glove boxes', a line of technicians and geologists numbered and described each lump of rock, weighed it, photographed it and took chips for close examination. A committee of experts sat in judgement on the samples, to decide how they should be divided among the groups in 200 laboratories around the world which examine the lunar material with every relevant technique known to science. At Houston the main concern was that the samples should reach those far-flung laboratories in virtually the same condition as they had been when lying undisturbed on the Moon's surface.

If the US space administration had communicated to the public the excitement felt among earth scientists about the lunar samples, it might have retained more political support for its manned flights. The redeeming result of the *Apollo* extravaganza has been the gathering of rocks from chosen parts of the Moon by intelligent test pilots (after a crash course in geology) and their safe return to Earth. The Moon samples may help to settle fundamental issues about the early history of our own planet.

For whatever reason, vast eruptions of lava, which made the big dark patches of the Moon, the maria, occurred between −3600 and −3000 MY. Since then, on its surface at least, the Moon has been almost a dead planet, unaltered except by a solar wind of atomic particles and the sandblasting of meteorites. One unexpected discovery has come, not from the samples, but from a magnetic instrument left on the lunar surface by the crew of *Apollo 12*. The instrument showed how the Moon responded to magnetic 'gusts' in the wind of atomic particles that blows from the Sun. It indicated the existence of a layer about 150 miles deep in the Moon in which electric currents were set up by the magnetic puffs. This zone may well have been the source of the Moon's activity, which made the maria. But preliminary measurements of heat flowing out of the Moon, made during the *Apollo 15* mission, contradict this result, by suggesting that the Moon is quite hot throughout.

The Moon is thought to be about 4600 million years old, roughly the same age as the Earth. The dark lava of the maria is akin to the ocean floors of the Earth, though much older. The light-coloured highland areas are older still. Like the continents on Earth, the highland rocks are richer in aluminium, which makes the rocks less dense. Among these highland rocks is anorthosite, a crystalline material that occurs occasionally in old assemblies of rocks on Earth.

When Scott, the commander of *Apollo 15*, reported excitedly to Houston, 'I think we've found what we came for,' he was looking at a half-pound lump of almost pure anorthosite. It was milky white and resplendent with crystals and it came to be called the Genesis rock, or number 15415 in the Houston catalogue of lunar samples. Geologists on Earth confirmed Scott's identification of the material and, although the first measurements indicated an age of 4150 million years, it may well be considerably older.

In the months that followed Scott's report of large quantities of anorthosite on the Moon, an opinion grew stronger among many earth scientists that the Moon began with a continent-like skin, rich in anorthosite, into which the lava of the maria burst 1000 million years later. At the same time, two contrasting views of the comparison between the Earth and the Moon became clearer. The more orthodox view is that the Moon was utterly different from the early Earth, which supposedly started with an ocean-like skin of basaltic rock, with little continental material present. The implications for earth science of lunar research would then be indirect, leading perhaps to an explanation of why the smaller body should have evolved quite differently from the Earth.

The other view is bolder and simpler. In it, the Moon is a more direct demonstration of what our own planet was like 3000 million years ago – a model of the Earth's past, differing in important details and less complicated than the Earth ever was, but correct in its main features. According to those who prefer this concept, the Earth began with a granite-like skin. Anorthosite was formed beneath it and the skin was broken by the development of 'maria' – the ocean basins. The reason why this similarity is no longer obvious is that almost all the old material on Earth has been reworked repeatedly, right down to the present, while the Moon 'ran out of steam' long ago. The most recent discoveries about the oldest rocks on Earth lend support to this view. It is a matter to which I shall revert, in considering the origin of the Earth's continents.

Astronomers who are forced to see the Earth as just a pebble in space talk of rearrangements of the solar system, with planets losing or capturing moons, or comets colliding with the Earth. They speak of exploding stars, which could be very damaging to life. Our passive Moon may record such events in its surface layers.

Samples have already disposed of one longstanding question: the Moon did not come out of the Earth, as some theorists have suggested, in a great break-up of our planet in the distant past. The composition of the Moon is sufficiently distinctive to rule that out. But did the Moon ever approach very close to the Earth? If so, huge tides, not just in the Earth's oceans but in the solid rock of both bodies, could have caused dramatic events in Earth and Moon. Even today, there are moonquakes every month, caused by the Earth.

The best estimate of the theorists is that the Moon came closest to the Earth at about —3000 MY and was then revolving around the Earth once every 6·5 hours at a distance of only 12,000 miles (20,000 km). If so, one effect would have been a temporary drying-out of the Earth's oceans. It may be significant that the last period in which large-scale melting of rocks occurred on the Moon was around —3000 MY, on the evidence of the lunar samples.

More definitely, craters caused by meteorites, which are conspicuous features of the Moon, also occur on Earth, although most of them are probably hidden. One crater 27 miles across, the largest in Europe, was identified in Sweden only by careful study. The heat and pressure of big meteorite impacts modifies the rocks locally. The Swedish crater formed at about —400 MY, but presumably large meteorites were much commoner in the early stages of the solar system than they are now. Before the end of the *Apollo* programme, it should be possible to date the formation of the very large craters on the Moon, such as Copernicus or Tycho, which may have been made by meteorites.

Important uncertainties

Did the Earth start hot or cold? If the raw material were mustered sufficiently rapidly, in a frenzy of meteorite-like collisions, the heat retained from the impacts would have given the young planet the thermal capital for operating its inner machinery from the outset. If, on the other hand, the formation of the Earth proceeded slowly, cooling may have occurred after each impact. There would then have been a long pause after completion while the planet generated, from its own resources of radioactivity, sufficient heat to melt rock.

The chief argument for a hot start lies in the evidence that the machinery was working quite early after the supposed formation of the Earth at —4600 MY. Ancient rock formations show there was mud – and therefore water – on the Earth's surface within 1000 million years of that date. Rocks were being magnetised when they formed, at least by —3000 MY, implying the existence of a molten core inside the Earth. It is hard to see how, from a cold start, the radioactivity could have produced big effects so soon. One escape for those who favoured a cold start was to say the Earth was really a good deal older.

The age of the Earth is normally said to be 4600 million years, although often the figure is quoted without thought as to what it means. The way it is arrived at is by measuring the ages of meteorites, material that originated outside the Earth and is very old. The relative amounts of lead atoms of different weights give a figure of 4600 million years. The second step is to show that the younger lead found on Earth has the atomic composition that would be expected if the Earth's original lead were the same as the meteorites'. By this rather roundabout argument, the Earth is judged to be the same age as the meteorites. But there is sufficient uncertainty to allow the Earth to be a few hundred million years older – and hence, perhaps, for it to have started cold and taken time to warm itself up.

Meteorites figure in another important way in arguments about the Earth. Before the first samples were brought back from the Moon, only two outside sources gave information about the raw materials from which

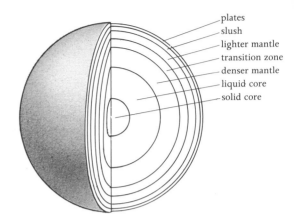

The interior of the Earth as revealed by past and present studies of how earthquake waves travel.

plates
slush
lighter mantle
transition zone
denser mantle
liquid core
solid core

the Earth was formed. One was the Sun. Careful analysis of sunlight reveals the presence of the chemical elements in the Sun's atmosphere. When the Sun's superabundance of hydrogen and other light elements is discounted, there is rough accord between elements common on Earth and elements common on the Sun. While the Sun is plainly quite different from the Earth, the other 'cosmic' source of information, the meteorites, are stony or metallic, like materials known on Earth.

Chemical analysis again reveals broad similarities between the stony meteorites and terrestrial rocks, but also some notable differences. The most striking are that the Earth's surface rocks have proportionately much more aluminium, potassium and titanium than the meteorites have, while sulphur seems to be curiously scarce at the surface of Earth. Weight for weight, stony meteorites possess seven times as much sulphur as the known rocks of the Earth; compared with the Sun, the Earth's deficit appears to be even greater. This curious scarcity of sulphur bears closely on the argument about a hot versus a cold origin of the Earth.

All earth scientists take it for granted that the main ingredient of the liquid core of our planet is molten iron, with a little nickel besides. The snag is that the core is not quite heavy enough; there must be some lighter material mixed in with the iron-nickel. In the 1960s, the favoured candidate was silicon. But silicon is not easy to separate from the oxygen atoms to which it is commonly bound in rock. Accordingly, Ted Ringwood in Canberra has described a hot start to the Earth, in which carbon took the oxygen from the silicon and itself disappeared into space as carbon monoxide gas. In this blast furnace, the Earth would have also lost much of its sulphur.

In 1970, another possibility was being urged. According to V. Rama Murthy and H. T. Hall of the University of Minnesota, the Earth did not put silicon into the core and throw out its sulphur; instead it put the sulphur into the iron core. Some colleagues seized upon this idea with relief and satisfaction because the Earth became much easier to make. Besides 'finding' its missing sulphur, as the lightening ingredient of the core, it also allowed the core to melt at a much lower temperature.

For John Lewis of the Massachusetts Institute of Technology, sulphur in the core opened another possibility: namely that sulphur going into the core could have taken potassium with it. The importance of potassium, for the machinery of the Earth, is that it is slightly radioactive. If much of the Earth's potassium were held in the core, it would give a big additional source of heat, both for stirring the core itself to generate the Earth's magnetism and for keeping the planet as a whole in a lively state.

Ringwood retorts that the amount of sulphur in the iron of meteorites is very small, yet the Earth's core would have to consist of about one quarter sulphur to pull its density down by the necessary amount. The pro-silicon and pro-sulphur factions are both engaged in laboratory experiments to test the behaviour of sulphur and iron combinations.

Controversies of this kind give a fair picture of the state of our knowledge about the origin and composition of our planet. There are important uncertainties, which have to be resolved, but they are quite specific, clearly stated and amenable to further research. Considering how inaccessible the Earth's interior is, and how long a time has passed since the Earth's formation, the range of possibilities has already narrowed remarkably. Much of the credit must go to the seismologists. By identifying the internal layers of the planet, they have filled in the 'down' words in the terrestrial crossword puzzle, leaving the 'across' words to be completed by the investigations of the geochemists.

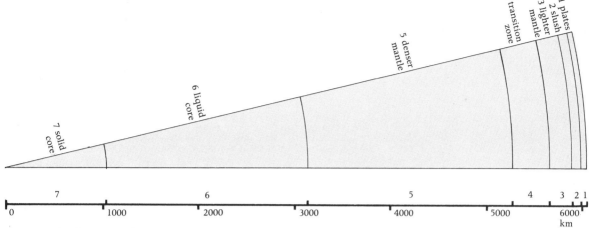

Layers within layers

The deep interior of the Earth reveals itself to us chiefly by its effects on earthquake waves passing through it. This exploration is like peering into a misty glass ball that is lit by intermittent flashes at its surface, which cast shadows and reflections of its inner structure. Recent progress in seismology has not been confined to the creation of better networks and the construction of large-aperture arrays described earlier (p. 63).

A major discovery was first hinted at in 1952 by Hugo Benioff, an outstanding figure in American seismology, and was then amply confirmed by Benioff and other seismologists at the time of the disastrous Chilean earthquake of May 1960. It turned out that an earthquake could set the planet vibrating, not only by the ordinary passage of seismic waves but by making the whole Earth ring like a bell. On that occasion it went on ringing for a month.

Just as any bell has its own tone, so the Earth has characteristic rates of vibration, known in the jargon as normal-mode oscillations. They are so slow that ordinary seismographs do not record them. One type of instrument, developed by Benioff, which can detect the normal-mode movements, consists of a long quartz rod projecting from one pillar and reaching very close to another; an electronic system measures changes in the gap between the end of the rod and the second pillar, and thus shows how the ground between the pillars stretches or contracts. Nowadays, laser beams are being used for this purpose of measuring 'earth strain'.

The slowest of the normal-mode oscillations occurs once every 54 minutes; the periods of others range down to just a few minutes. The importance of the discovery is that it provides a new way of deducing the construction of the Earth as a whole, much as the sound of a bell might tell an expert what kind of bell it was, even if he could not see it.

Without trying to mention all the advances that have been made in seismology in recent years, I should emphasise the role of the computer, which has helped to transform human understanding of the deep interior of the planet. It assists in disentangling complex seismic records. It also makes possible very elaborate calculations, for example those concerning the 'ringing' of the Earth; the computer allows the invention and testing of thousands of possible descriptions of the Earth's structure, to see which best fits the seismic observations.

Important features of the Earth's interior were discovered by the pioneer seismologists before 1914. They divided the Earth into layers: the *crust*, through which seismic waves travel quite slowly, the *mantle*, where they go much faster, and the *core*, which is molten, as proved by its inability to transmit certain kinds of tremors. The boundary between the crust and the mantle, where the waves suddenly change speed, is called the *moho* as a convenient abbreviation for its Croatian discoverer, Mohorovičić. Another major feature of the Earth revealed itself in 1936; a solid or semi-molten *inner core* at the very centre of the Earth.

The recent work in 'X-raying' the Earth with the aid of seismic waves has not marred the old picture but it adds accuracy and significant new detail. The waves gather speed progressively as they go deeper into the mantle; this is only to be expected as the density of the material of the Earth increases under the enormous weight of overlying rock. But now additional layers are known, deep within the Earth, where the speeds change quite sharply.

The crust is relegated to a minor status in the general structure of the planet. Instead, it is better to think of the mobile *plates* as the main outer layer of the Earth. At about 40 miles in thickness, the plates are much more substantial than the pieces of the crust that they carry, which are typically 3 miles thick under the oceans and 22 miles on the continents. Between the base of the crust and the base of the plate, the rock transmits seismic waves fairly rapidly.

Beneath the plates, the waves slow down appreciably, in what is therefore known as the low-velocity zone. For simplicity, I shall call it the *slush* because that is what it almost certainly is: solid rock mixed with a certain amount of molten rock.

Below the slush, the mantle is solid all the way to the core, but there is a *transition zone* between 250 and 500 miles deep in which the seismic waves gather speed more rapidly than they would do if the rocks were simply more compressed. Under the very high pressures of overlying rock, the atoms arrange themselves differently, so that they take up less space. In other words, the crystal structures in the rock are not the same as those adopted higher up. A major preoccupation of investigators of the Earth's interior is the search for mixtures of materials that would behave appropriately at the various depths.

The olivine planet

At the Australian National University in Canberra, Ted Ringwood and his colleagues operate powerful presses that can squeeze crystals to pressures of more than a thousand tons to the square inch. They use them for imitating conditions that prevail hundreds of miles below the surface of the Earth. Research of this kind, in Canberra and elsewhere during the past ten years, has helped to achieve what seems at first sight impossible: to explore the composition of the deep interior of our planet, where no one could travel and no borehole is ever likely to reach.

The main ingredient of the Earth is olivine, or olivine transformed under high pressure. This material is known as a green gemstone; it is a compound of iron, magnesium, silicon and oxygen, which are by far the most abundant elements in the Earth. There were strong hints that olivine might be the key material: it is a suitably dense substance, it occurs abundantly in many meteorites and it is found at the Earth's surface wherever very deep-lying rocks have forced their way up.

Experiments at high pressure clinch the case for olivine. They show that important transitions in the internal state of the Earth, detected by earthquake waves, are very readily explained. Ringwood and other investigators find that olivine crystals collapse under pressure to make denser crystals, in just the conditions matching those expected 250 miles down, where the transition zone begins. Conditions corresponding to much greater depths cannot yet be achieved in laboratory presses, but there are ways around this difficulty. One, employed in Canberra, is to test similar crystals that change form at lower pressures. Another is to use guns or various explosive devices to subject materials to extremely high pressures, even though only for an instant.

There is no longer any reason to doubt that the rocky interior of the Earth consists mainly of olivine, mixed with less dense materials of other chemical composition. The precise mixture is still a matter for debate but it undoubtedly includes garnet (rich in aluminium) and pyroxenes (rich in silicon). At very great depths these ingredients, as well as the olivine, are transformed into more compact materials. On the other hand, when hot rock from the interior finds its way up to the surface of the Earth, the crystals 'relax', as the pressure diminishes, and change into less dense forms.

The stuff that comes right to the Earth's surface at

Volcano in New Zealand. The northern end of New Zealand is one of several regions where the Pacific plate is being destroyed.

Icelandic rocks, representing an early stage in the evolution of the minerals of the Earth's surface. Much of the basalt of the ocean floor, here exposed to view, is eventually returned into the Earth at ocean trenches. Some of it is reprocessed at the trenches, to make more durable continental rocks.

the mid-ocean ridges, to make the floors of the growing oceans, is simply material from the interior that has partly melted. It has left behind some of the heavy olivine, the separation possibly starting at a depth of about 30 miles. The denser residue normally comes into daylight only at the scene of drastic events – in mountains, where the surface has been buckled and squeezed, or in diamond pipes, where deep explosions have punched holes through the Earth's outer shell.

Michael O'Hara of Edinburgh University dissents to the extent of asserting that the unmixing of the ingredients of the interior which are ascending towards the surface begins at a much deeper level. In his view, some of the liquids so formed make eclogite, a material found in diamond pipes and in the roots of mountains. If that were so, the molten material reaching the surface would be less closely related to the rocks of the deep interior, and present deductions about details of the Earth's composition would be put in question.

Most rocks of the continents are very different from olivine, so it is natural to ask why the outward appearance of the planet is so thoroughly misleading about its internal composition. One answer is that, if the continents were not very different from the interior, they would not exist. To survive, they have to be light enough to remain afloat on the denser underlying rocks, even when the plates on which they are riding dive back into the interior of the Earth at the ocean trenches.

A more constructive explanation is that any volcanic or other process which allows the lighter components of the interior to break loose will tend to create a kind of buoyant scum on the Earth's surface. In practice, a two-stage process seems nowadays to be necessary for making new raw material for continents. The mid-ocean ridges manufacture ocean floor that is lighter than the interior rocks from which it derives, but not

Within an old rock province in Rhodesia a crowd of granite domes (contoured) occurs in a 'sea' of greenstone belts (black). (Adapted from map of A. Holmes.) The photograph below shows one of these Rhodesian granite domes. The 'bowler hat' is a superficial feature on a much bigger dome.

yet light enough to avoid recapture at the ocean trenches. As it goes down, however, the friction of the oceanic plate promotes volcanic action of the island-arc variety which produces material light enough to stay up for ever.

The crucial result of this second stage is to make rocks that are rich in silicon and aluminium. These elements, combined with oxygen, form crystals that are appreciably less dense than the iron and magnesium compounds of the ocean floor and the Earth's interior. The continents then evolve and make materials as various as rubies and coal. During mountain-building further heating occurs deep in the continental material, and granite is manufactured, which many regard as the quintessence of the continents.

These processes are going on now and the continents continue to grow. New continental material is forming in the Andes and in Japan and other island arcs, while consolidation of older continental material proceeds in the Himalayas and the Alps. But did the continents start in this way?

The greenstone enigma

'Like granite bubbles blown through the earth' – that was Lawrence of Arabia's description of hills he encountered near the Red Sea. Of one of them he added: 'seen from nearby, it more resembled a huge football half-buried in the ground.'

Round domes of hard rock, surmounted by natural 'bubbles', occur in some of the oldest parts of the Earth's land surface and they may have something significant to say about the origin of the continents. The best-studied collection of granite domes is in Rhodesia, where nearly twenty of them stand clustered in a group. As Christopher Talbot of Dundee University describes them, they vary in diameter from 15 to 120 miles, with gaps between them in the same range of distances. They could have arisen in the breakdown of a thin, heavy layer of rock overlying a comparatively light layer. The lighter layer, being buoyant, could have penetrated the upper layer at many places, as a swarm of domes.

To Talbot that seemed at first the obvious explanation. But there are two puzzling aspects. One is that granite normally forms in strips, like the Sierra Nevada in California or the line of domes in south-west England (see page 45); the Rhodesian clustering is peculiar. The other enigma is the material that lies around and among the granite domes: a type of flaky rock known as greenstone. Belts of greenstone elsewhere have been interpreted as the relics of the floors of very old oceans. Talbot has reluctantly shifted to the idea that these granite domes, far from being the products of a multiple birth, were microcontinents that formed separately and drifted together.

The association of granite and greenstone occurs over large areas of the oldest parts of the continents, notably in South Africa, in Canada and in Western Australia. Let me outline one coherent theory of the origin of the continents and then mention some awkward questions about it. Present-day island arcs, such as Japan, may illustrate the way in which the continents began.

Suppose that the Earth starts with oceanic plates, which are already somewhat different in composition from the interior rock of the Earth. One plate dives down under another and generates the volcanoes of an island arc. The Earth's material is thereby reprocessed and the part of it that erupts in the volcanoes is light enough to float, whatever happens. This is the raw stuff of continents. Thereafter, the material is reprocessed by erosion and by heating, to make granites and other characteristic continental rocks. The island arcs coalesce to make sizeable continents, trapping between them the oceanic material of the greenstone belts.

Now for the queries. Was the Earth cool enough, when the continents were first being formed, for plates to be rigid enough to make island arcs by frictional heating? How does it come about that one province of ancient rock (the Superior province of Canada for example) can be much wider than a typical island arc? Again, island arcs have comparatively thin rations of crust, so did the old continental masses start off as thick as they are now, or have they grown thicker by later addition of material to their undersides? And here we return to the oldest known rocks on Earth, those found by Vic McGregor in Greenland. They are granitic rather than volcanic. Does that not suggest at least that some continental material may have arisen without the aid of island arcs?

The sense of all this cross-examination of the island-arc theory of continental origin is that there is an alternative story. It tells of a skin of granite-like rocks made before the oldest known volcanic rock, at a very early stage in the history of the Earth, and later broken

Kilauea erupts on the island of Hawaii, in the heart of the Pacific plate. According to one theory, it stands over a 'hot spot'. Alternatively, the Earth's shell is 'leaky' on an old fault line.

Contrasting ideas about the origin of continents. In 1 collisions of island arcs amass continental material. In 2 the breaking of a primeval granitic shell forms continents.

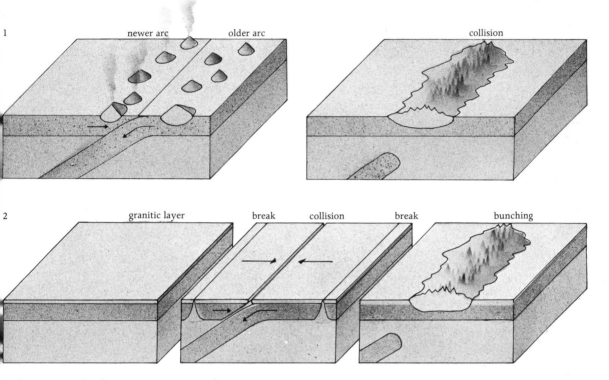

and swept together by emergent oceans. The importance of the very old Greenland rocks, with which we opened this chapter, is that they tip the scales favourably, though not yet decisively, for this alternative view. The rocks are a leafy material (gneiss) that are themselves derived from still older granite, now lost. That parent granite may have been the Earth's primeval crust.

The uncertainties are still great enough for experts to hint darkly at astronomical events affecting the early Earth, including wholesale bombardment by huge meteorites, or great upheavals produced by the Moon. Any irritation about the lack of a definitive account of the beginnings of the continents we stand on would be misplaced. The exploration of the oldest parts of the crust is progressing faster than ever before, greatly aided by the improvement in dating techniques, and it is remarkable how much is being found out about early events. More disturbing is the fact that we cannot fully explain what is going on under our feet, now.

Hot spots and cold sinkers

The Hawaiian islanders tell how, with her magic spade, Pele the volcano goddess travelled to each one of their islands in turn, before settling where she lives now, in her palace of Kilauea on the big island of Hawaii. As myths go, this one is uncannily well borne out by atomic dating techniques. The Hawaiian islands emerged by volcanic eruption from the deep ocean floor, one by one, at intervals of roughly a million years. From the oldest island the chain trends towards the south-east, to Hawaii itself, the youngest island. Current eruptions occur in the south-east corner of Hawaii – Pele still has itchy feet.

Some conservative earth scientists talk of Hawaii as if the new theorists of plate movements really need that magic spade to explain it. Why should volcanoes pop up in the middle of an oceanic plate, where there is no rifting (as in Iceland), nor any plate swallowing (as in Japan), nor even any significant earthquake activity?

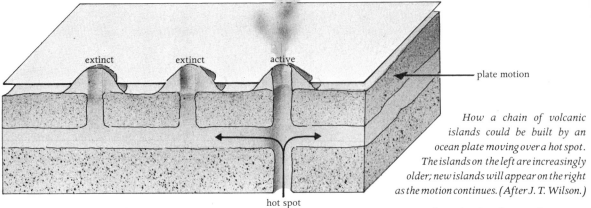

extinct   extinct   active

plate motion

hot spot

How a chain of volcanic islands could be built by an ocean plate moving over a hot spot. The islands on the left are increasingly older; new islands will appear on the right as the motion continues. (After J. T. Wilson.)

One of the founders of plate tectonics, Jason Morgan of Princeton, is undaunted. On the contrary, he has recently developed a theory that offers to account for Hawaii, to use such island chains to track past plate motions, and to provide the driving force that moves the plates about. For Morgan, pursuing an idea of Tuzo Wilson's, Hawaii is the product of a hot spot.

The Pacific plate on which the Hawaiian chain stands is moving north-west. According to Wilson and Morgan, a strong source of volcanic energy – a hot spot – lies more or less fixed in the main body of the Earth. Like a worker at a conveyor belt, it repeatedly punches through the plate moving above it, producing exactly the line of islands that now exists. Other parallel lines of islands and submerged seamounts exist further south on the same Pacific plate, made by other hot-spots: the Tuamotu and Austral Islands.

But all three lines eventually turn, in their oldest parts, and run in a more northerly direction. For example, the Austral Islands link up with the Gilbert and Ellice Islands. The reasoning can now work the other way. If the present motion of the plate over the hot-spots produces the younger islands, the trend of the older islands marks a period when the Pacific plate was moving more nearly northwards. Morgan estimates the change of course at 40 million years ago. The full extent of the chains gives, he thinks, a record of 100 million years of Pacific plate history.

There are many other candidates for hot spots, most of them associated with shallow parts of the ocean that are volcanic but essentially free from earthquakes. In the Atlantic, Iceland is, on this theory, a prominent hot spot: so are several other islands all the way to Tristan da Cunha and beyond. If Iceland is over a hot spot, then the volcanic islands of Scotland also were made by it. In the USA, Yellowstone is said to be over a hot spot, but otherwise Morgan nominates very few hot spots under continents. He suggests, though, that former hot spots will have left their marks both on the ocean floor and on the continents.

He takes the idea further. Hot spots, according to Morgan, are the outlets of pipes that carry up hot material from deep inside the Earth. When it nears the surface, the hot material spreads horizontally in all directions away from each hot spot, before sinking back into the Earth. This movement away from the hot spot helps to propel nearby plates, although the actual motion of a plate depends on its interactions with other plates and other hot spots. Morgan estimates about twenty hot spots dotted around the world.

The question of what moves the plates of the Earth's outer shell is high on the agenda of research. The plates and their interactions constitute a fine piece of machinery that explains the outward appearance of the planet. It is exasperating not to be sure how the machinery works, the more so because we already know what the fuel is that energises the movements – radioactivity in the rocks.

Are Morgan's twenty hot spots the rocky motor of the Earth? Unfortunately there are other mechanisms on offer for moving the plates of the Earth. One is that the material rising at the mid-ocean ridges forces the plates apart, like a never-ending wedge. Another is that the piling up of material at the mid-ocean ridges allows the plates to slide downhill under gravity. Again, a cold, sinking slab at an ocean trench may drag towards the trench the rest of the plate to which it is attached. Closely related to this idea of the cold sinker is the notion that, after a certain period of cooling on the ocean bed, the old portion of a plate becomes 'tired', as it were; it grows dense enough to sink spontaneously. There are arguments for, and more often against, each of these mechanisms as a realistic and sufficient explanation of the plate movements.

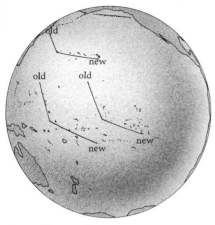

Mid-ocean submerged sea mounts have been built along three trac... the Pacific during the past 100 million years. According to Jason Morgan, they are the product of three hot spots, over which the Pacific plate has moved. The 'elbows' correspond to a change in direction of motion of the plate, 40 – 50 million years ago. A partial track lies near Alaska. (After W. J. Morgan.)

Various suggestions about what propels the plates of the Earth's shell. (1) Hot spots: plates tend to move away from them.
(2) Push from the mid-ocean ridge: plates move 'downhill'.
(3) Tug from the trench: plates pulled by sinking part.
(4) General convection: hot rock rises, cold rock sinks.

Caterpillar tracks

The hot-spot theory sets the pioneers of plate tectonics at odds. Dan McKenzie challenges Morgan's notion that certain parts of the interior of the Earth are unusual. The lines of volcanic islands certainly exist but, until there is stronger evidence to the contrary, McKenzie prefers to suppose that the lines of oceanic islands occur along old faults in the ocean plates. These faults provide cracks of weakness through which material can leak from the Earth's layer of slush. A line of islands in the Indian Ocean ranging north from Mauritius is made that way. So, too, McKenzie would say, is one of Morgan's hot-spot tracks, the Emperor Islands of the Pacific, where the piece of ocean floor on one side of the line of islands is much older than that on the opposite side.

Of the other ideas about what propels the plates, McKenzie takes the view that they are just different aspects of one story. The story is the circulation of the material of the Earth by convection – that is to say, by hot material rising towards the surface, spreading sideways and cooling, and then sinking again. It is roughly the same kind of motion as occurs in a saucepan of water over a flame. Heating makes a portion of water rise, because it becomes less dense than the surrounding water; then, cooling at the surface makes it denser and it drops back to the bottom. If it is difficult to visualise solid rock behaving in this way, moving very slowly and viscously, it is even harder to calculate how the motion proceeds.

One reason is that, unlike water in a saucepan, which is heated from below, the rock carries its own source of heat, in the form of radioactive material. Rocks also flow more readily when they are hot than when they are cool. One distinguished authority, Arthur Holmes, advocated in the 1930s the possibility of convection in the Earth; another, Harold Jeffreys, declared it to be

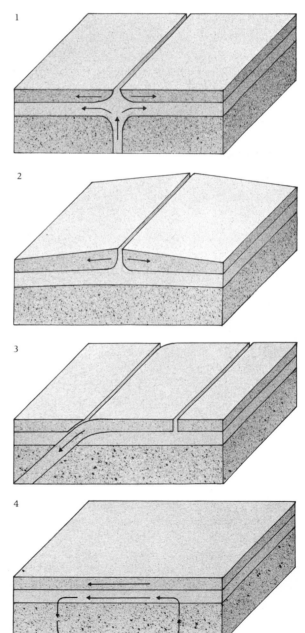

When a fluid is heated from below, it usually forms roughly square convection 'cells', as above. When it is heated from within (as the rocks of the Earth are, by radioactivity) the 'cells' can be much flatter.

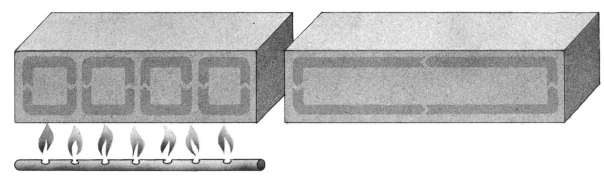

impossible and was still saying so in 1970. But in recent years the younger generation of earth scientists has been inclined to accept convection as the main explanation for continental drift and the movement of plates. Several groups have made theoretical studies of how the convection proceeds.

Theorists are now embarked on a series of calculations by computer to find patterns that may be set up within the Earth by such a wheeling motion driven by heat. The 'cells' of upward, sideways and downward flow of rock are confined, in some definitions of the process, to the uppermost 440 miles of the Earth.

The material of the Earth possibly moves around elongated loops, like the caterpillar tracks of a tank. The plates at the Earth's surface, in this description, are simply the coolest and most rigid part of the material taking part in the sideways motion at the top of the convection cells. Beneath the plates, and supposedly moving in the same direction, is a layer which scarcely has the chance to cool and is indeed close to melting. Here is the slush, or low-velocity zone, which we encountered earlier (p. 100). If any crack appears anywhere in a plate, as at a mid-ocean ridge, it is immediately filled by material welling up from that layer. The ocean trenches mark regions in which cold material falls back into the Earth's interior.

Some of the other mechanisms already mentioned, as possible agents for moving the plates, take part in the heaped-up material of the mid-ocean ridges, and the dragging effect of the cold sinkers are not, in McKenzie's view, adequate explanations of plate movement but they contribute to the general motion and have to be taken into account.

If convection cells exist within our planet, there is no clear picture of them. The downward sinking cold slabs give themselves away at great depths, both by the earthquakes they cause and by their effect on earth-quake waves passing through them. Perhaps the extremely sensitive seismic instruments now available will also pick out rising masses of hot material. If Morgan is right and narrow pipes of rising material exist, these may cast noticeable shadows in the seismic view of the Earth's interior.

The map of the Earth gives no sense of a regular upflow and downflow. The mid-ocean ridges and the ocean trenches seem to arrange themselves almost randomly about the Earth; moreover they are broken and they move, so they can hardly be tied to columns of rock running deep into the Earth. Perhaps the motions are opportunistic, with heat escaping from the interior where it can and cold material returning where it must. But the disposition of mountain chains also appeared meaningless until continental motions very recently explained it. There may well be a pattern of convection and plate motion that has still to be spotted.

Although no one knows exactly how the plates of the Earth's shell are driven around, that does not deny their movement. No one doubted that the Sun shone, though astronomers had to wait thousands of years before learning that it burned by nuclear energy. The plates have moved, and they have shaped not only the rocks of the Earth but the patterns of life upon it.

Mount Fuji, Japan's most famous volcano. It consists of comparatively light rocks that erupt at the surface as the Pacific plate dives to destruction underneath Japan. This action enlarges the land but it is accompanied by deep-seated earthquakes.

Chapter 5　Living on Rafts

The Earth would be a quite different planet without the thin film of life on its surface, but life, in turn, has been profoundly affected by the motions of plates and continents. These motions are helping to explain why marked changes in living conditions have occurred, with identifiable effects on evolution.

Why did the dinosaurs die? That is the representative question about the abrupt changes in the species populating the planet, as revealed in the record of life that is fossilised in the rocks. When millions of years of Earth history are reduced to a few feet of rock, changes can seem more sudden and dramatic than they really were. Nevertheless, the great reptiles which ruled the Earth for 150 million years are no more – or at least they are represented by nothing more impressive than a crocodile. Those Hollywood productions which show prehistoric men battling with rubber tyrannosaurs are, of course, misleading; no men were around to hasten the end of the great reptiles seventy million years ago.

Attempts to explain the extinction of the dinosaurs have not been lacking. The choice ranges from speculation that artful little mammals ate the big reptiles' eggs to the notion that a comet collided with the Earth. There is no authoritative answer to the question, yet consideration of it, and of related issues about life on Earth, will lead us to explanations for other catastrophes. It will also help to show a remarkable interplay between life and the physical and chemical processes of the planet.

Whatever it was that spelt death for the dinosaurs also axed the ammonites. At one time these jet-propelled sea animals were about as common as fish are today and their shells, typically coiled, are the best known of all fossils. So abundant are they in the cliffs of Yorkshire that local legend attributes them to a plague of snakes quelled by St Hilda. They have been completely extinct since —70 MY, like the dinosaurs. But the ammonites had undergone extraordinary reversals of fortune

White cliffs of south-east England, at Birling Gap, Sussex. This chalk consists mainly of the remains of living organisms which accumulated on the bed of a shallow sea, about 100 million years ago. The Alpine collision of Italy with Europe later pushed the English chalk into the air.

during the previous 300 million years in which they roamed the seas.

Three times before their final demise they were reduced near to extinction; three times they made a spectacular comeback. In those intervening periods, life must have been very easy indeed for the ammonites. Their evolution was riotous and many stocks prospered even when they developed bizarre and unwound shells in place of the trimly coiled shells that had served them well in difficult times. Indeed, the ammonites seem to have wandered repeatedly and once too often into evolutionary traps; when conditions worsened, the fancy forms were in trouble. The new geology begins to give explanations of what those changes of conditions were.

Compared with the thousands of miles of rock beneath our feet, life is little more than a patchy film covering the planet. It is influenced by the continual changes in the rocky surface, brought about by plate action and the ponderous forces of the Earth's interior. The movements of the continental rafts on which we live have greatly affected the course of biological evolution.

Yet living things have certainly not been mere passengers on the restless Earth. The green pigments of the plants, whether giant trees or microscopic plants floating in the oceans, trap altogether about a thousandth part of the energy of the Sun's rays falling on the Earth. It is sufficient to grow 150 billion tons of new living material each year – more than twice the rate at which volcanoes and other upwelling rocks add new material to the crust of the Earth. Most living tissue is, of course, less durable than rock but all of it affects the chemistry of the planet's surface and that which is preserved has helped to make minerals useful to man, as well as contributing its multitudes of corpses to the manufacture of mountains.

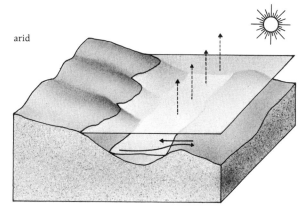

arid

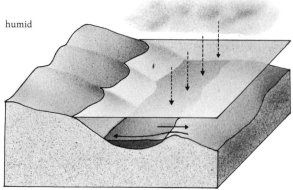

humid

The crumbs of Herodotus

The diminutive animal, *Heterostegina*, which lives in a coiled shell on the sea bed, is impressive in its self-sufficiency. It requires nothing more for life than sea water and sunlight, because it is one of several sea animals that maintain their own kitchen gardens. Inside the semi-transparent whorl of *Heterostegina* you can make out greenish patches that are algae, microscopic plants to which the animal affords shelter at the same time as it feeds upon them.

Rudolf Röttger, a young researcher at Kiel University in Germany, was first to observe the reproduction of *Heterostegina*, captive in his laboratory. As he described it to me, the event is sudden, with only a little leakage of protoplasm from the adult animal to give any warning; within a couple of hours it is over and the adult is dead, leaving only a blanched shell of calcium carbonate, the stuff of limestone. In the process, the parent expels into the surrounding water hundreds of offspring; each of them is endowed with a transplanted share of the parent's algal garden and the life cycle begins again. In six months or so the shell of the *Heterostegina* will add many more chambers to its shell than the two with which it is born; then it, too, will be ready to reproduce.

The dead shells of the adults accumulate on the sea bottom and make new limestone rock. In the Persian Gulf, German marine geologists have found *Heterostegina* adding half an inch per century to the limestone forming on the sea bed. An ancient visitor to the Great Pyramid of Egypt, the Greek historian Herodotus, reported that the ground around it was littered with what he took to be remains of the food eaten by the slaves who built the Pyramids. They were in fact fossils closely akin to *Heterostegina* which had been weathered out of the stones the Egyptian stonemasons had chosen for the big structure. The limestone had been made

about fifty million years ago in conditions like those prevailing today below the shoreline of the Persian Gulf.

The animals that are the most effective in making rock are those with hard shells that also grow and reproduce with unusual productivity. *Heterostegina* is only one of the more notable of the 'foraminifera', so called because of the pores in their shells. Other productive kinds of rock-building animals include the corals and extinct forms of sea-lilies. The white cliffs of Dover are a part of a huge chalky accumulation of very small plants that lived in a clear, shallow sea formerly covering much of Europe. *Heterostegina* sometimes shows irregularities in the shells, apparently resulting from changes in temperature or salt content of the water. If so, irregularities found in fossil shells may help to define more precisely the environment in which the owners of the shells were living. At Kiel and in other marine laboratories around the world, this reconstruction of environments of the distant past is an objective which bears closely on attempts to assess the present health of the world's oceans.

A guiding theme is the division of the seas and oceans of the planet into two main types. Eugen Seibold, leader of marine geology at Kiel, picks out two 'models', the Baltic Sea and the Persian Gulf, which are important regions for current research. The one is a humid and the other an arid sea. Into the Baltic, rain and rivers bring abundant fresh water, rich in nutrients, which flows out of the sea at the surface while salt water flows in from the ocean at greater depths. In the Persian Gulf the situation is reversed. High evaporation makes the sea unusually salty and the outflow runs deep, while less salty water flows in from the ocean at the surface. That ocean-surface water is poor in nutrients for supporting life.

The formation of limestone rock proceeds apace in

Two kinds of seas (left) Life and rock formation in a sea are both greatly influenced by the pattern of inflow and outflow of water. 'Arid' seas include the Persian Gulf, the Mediterranean and the Atlantic; 'humid' seas include the Baltic and the Pacific. (After E. Seibold.)

Fossil foraminifera, small sea animals (below) seen in a scanning electron microscope, at magnifications of 150 to 400. They show a great variety of forms, recognition of which is an important aid in determining the age of marine deposits.

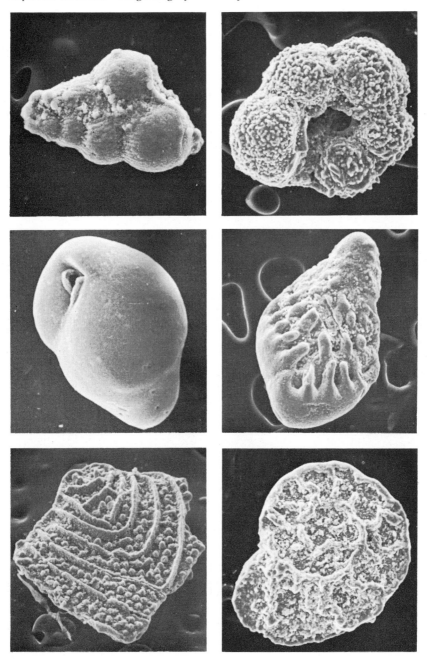

Coal in Antarctica. Black seams of coal are visible in the cliff face. They were laid down about 250 million years ago, when Antarctica lay at a more temperate latitude.

An ammonite that lived in the sea about 150 million years ago. This kind of animal survived, with varying fortunes, for hundreds of millions of years, but became extinct 70 million years ago.

the Persian Gulf. In the Baltic Sea, by contrast, the fresher and slightly acid water dissolves the carbonate shells of the animals soon after they die. Instead, rich black mud collects on the bottom of the Baltic, contrasting with the grey mud, depleted in organic matter, that forms on the bottom of an arid sea.

The Baltic provides a model for the Norwegian fjords and the Black Sea, while the Mediterranean is in the same arid class as the Persian Gulf. One hypothetical illustration of the importance of these circulation patterns is, as Wolfgang Berger of Kiel has suggested, that the underwater ridge, or sill, at the mouth of the Mediterranean should be blown off to let the ocean waters come in at a lower level, with a view to making the Mediterranean more productive of fish.

Berger extends the classification of sea types to include the oceans. The Atlantic, he says, is essentially of the arid type, being relatively salty and well oxygenated, but poor in nutrients and forming large quantities of lime-rich oozes on its bed. The Pacific is the reverse; it is better supplied with nutrients in its upper waters and forms lime-rich oozes much less readily. The typical ooze of the Pacific contains plant and animal remains rich in silica – material that, when it is consolidated into rock, makes chert instead of limestone.

Coal and oil are, next to the limestones, the most prominent mementoes of past life on this planet. They formed from organic remains that failed to decay in the normal way, because they were covered with stagnant water before they could do so. Peat forming in swamps represents an early stage of coal formation. To make thick layers of good black coal, like those on which the industrial ascendancy of Europe and the USA were founded, requires special conditions.

For the coal layers to be thick, you need to be growing the luxuriant vegetation of a tropical jungle on land which is slowly submerging; the sinking first allows one layer of vegetation to be drowned before it rots and then lets new forests spring up on top of it. To make the coal rich and black you then need the land to go on sinking so that it can be covered with a two or three miles' thickness of overlying rock. The pressure squeezes the water and gases out of the peaty material. These conditions were all fulfilled in the slow collision around —280 MY, between Africa on the one hand and the conjoined continents of North America and Europe on the other, when they were lying in the tropics.

Petroleum is not simply a marine equivalent of coal, although it usually requires for its formation a warm sea, the bottom of which has been depressed, creating stagnant, oxygen-depleted pools. Bacteria are required, to attack the corpses of marine plants and animals that drop into the pools; the bacteria tear out the oxygen atoms bound in organic molecules, leaving a residue of hydrocarbons. This process can be detected going on today in the bed of the Black Sea. It is slow: the highest layers of mud in which hydrocarbons are detectable are several thousand years old.

Like coal, though, petroleum has to be pressurised by a weight of rocks to become high-grade material. The overlying rocks must be impervious, otherwise the oil simply escapes. Being light and fluid, the oil tends to move sideways and upwards from its place of origin until it finds a trap – usually the highest point of an arched layer of pervious sand or else a place where the layer has been blocked by other material coming up from below. As domes of salt forcing their way upwards tend to block oil-bearing layers and form pockets where oil can collect, these salt domes are favourite targets for oil prospectors.

The Persian Gulf, probably the greatest of the world's oil fields, owes its origin to the slow crumpling of the sea floor by the collision between Arabia and Asia. Continental movements also explain why oil and natural

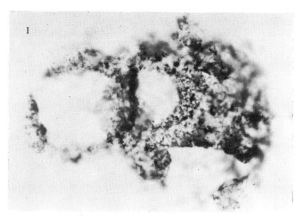

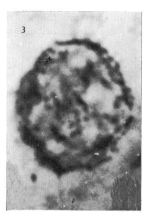

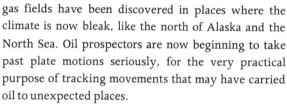

gas fields have been discovered in places where the climate is now bleak, like the north of Alaska and the North Sea. Oil prospectors are now beginning to take past plate motions seriously, for the very practical purpose of tracking movements that may have carried oil to unexpected places.

Out of the soup

The volcano called Oldoinyo Lengai lies in the Tanzanian end of the Great Rift Valley. The Masai call it 'the mountain of god' but he must be a stern god; the volcano's eruptions befoul the countryside with washing soda – sodium carbonate – which turns soapy when it rains and poisons the waterholes. Lengai last erupted in 1966 and it looked cool and peaceful enough when we peeped into it five years later. But Laurence Williams, a volcano expert from Nairobi University who had climbed Lengai at the time of the eruption, was with us in the light aircraft and he pointed out the unusual soda lava that daubed the cone like snow.

What looks like improbable tropical ice covers the nearby lakes, Natron and Magadi. Again it is washing soda, collected and recrystallised in the water of the lakes. Curious bloody streaks in the dazzling white crust of soda turn out to be populations of algae that thrive in this chemical environment. The soda is harvested commercially. At another lake, further north, we found an industry on a much smaller scale in the persons of two pedlars garlanded with lumps of bicarbonate of soda. They had scraped them from the edge of Lake Hannington and, when we met the pedlars, they were setting off to sell the grey material in the nearby villages for culinary and other, more mysterious, purposes.

The soda of the Rift Valley signifies the contributions that volcanoes have made to life on Earth. The water that fills the oceans and the nitrogen of the air, both necessary for life, probably originated in volcanic eruptions. So, too, did the carbon, the element with the versatile chemistry that makes life possible. More than any other type of volcanoes, those of the Rift Valley have carried the reprocessing of rock to the stage where carbon dioxide gas and carbonate lavas erupt in large quantities.

The Rift Valley and Oldoinyo Lengai are relatively new features of the Earth, but similar, slowly simmering chemical retorts were presumably at work in the Earth's youth, supplying the carbon from which the first living things could evolve. Chemists who consider the origin of life on Earth imagine a 'soup' forming as the water became enriched with carbon compounds of considerable complexity. Nor is this pure speculation, for laboratory experiments show that these simple raw materials and the action of the Sun's rays, and other stimuli, could not have failed to form vital chemicals – protein for example – in abundance.

The expectation from this imaginative treatment of the origin of life was that, first of all, quite organised little 'blobs' would concentrate materials from the soup. These 'blobs' would not be alive but eventually, by chance, a living combination of chemicals would emerge. The first organisms would be simple bacteria eating up the soup, a way of life that would be quickly self-defeating unless the organisms acquired a means of exploiting the energy of sunlight to make their own food, as plants do today.

In 1963, John Ramsay was investigating rock formations in South Africa when he discovered what appeared to be microscopic fossils in very old rocks, the so-called Fig Tree cherts, near the Swaziland border. Shortly afterwards a dedicated hunter of early life, Elso Barghoorn of Harvard University, collected rocks from the area. On examination, they proved to contain

The origins of life (left) are revealed by three kinds of microscopic objects found in South Africa. The 'blobs' in 1 (3400 million years old) are either the first living organisms or their immediate non-living precursors that floated in a chemical soup. The oldest known bacterium in 2 occurs in somewhat younger rocks, along with 3, the most primitive sorts of algae.

Eruption of Oldoinyo Lengai (below), the washing-soda volcano in Tanzania. Much of the carbon necessary for life on Earth may have come to the surface in volcanic activity of this chemical type, which owes its peculiarity to the comparatively slow rate of rifting in East Africa. Similar volcanoes may have been active when the planet was young.

A big, dome-like colony of fossil algae, called a stromatolite. It was alive with vast numbers of individual organisms, more than 1200 million years ago. This specimen was found in the old rocks in northern Canada.

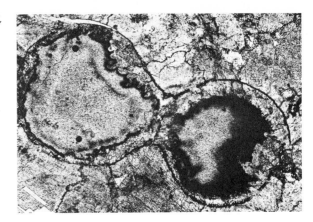

A small organism 2000 million years old, from Greenland, fossilised at the moment of reproducing itself.

rod-shaped bacteria quite similar to some existing bacteria, and also 'coalified' algae, apparently ancestors of the most primitive type of single-cell plants that survives today, the blue-green algae.

The Fig Tree rocks are at least 3200 million years old and seem to represent one of the very earliest stages of life on Earth. This supposition is supported by the discovery of 'blobs' in the same region, in rocks that are about 3400 million years old, the Onverwacht Series. These 'blobs' look quite like microbes but they vary much more in size – by a factor of 30 – than you would expect a well-formed organism to do. Besides the fossils, the rocks contain organic chemicals. It begins to look, therefore, as if the 'blobs' floating in the soup turned into bacteria and the first microscopic plants some time between —3400 and —3200 MY.

The rise of oxygen

Considering that life began so early in the planet's history, the events in the next 2000 million years seem at first sight unimpressive. For half its history the Earth was alive, but less so, by our standards, than a fishpond without fish. The old rocks show nothing more palpable than the lumps built by colonies of algae; most of the fossils have to be found with a microscope. Yet revolutionary innovations occurred in minute organisms during this long twilight before the dawn of conspicuous life – natural inventions that we take for granted.

Photosynthesis, the ability of plants to grow by sunlight, was essential almost from the outset. A later development was the capacity to use oxygen constructively, as a prelude to respiration. In those early aeons, oxygen was probably very scarce on Earth, but growth by sunlight produced oxygen in and around the primitive plant. Although it is usual nowadays to explain the rise of oxygen in the atmosphere by the action of green plants, an alternative theory accounts for the oxygen by the Sun's rays breaking up water vapour in the upper atmosphere. The hydrogen would wander off into space while the oxygen, being heavier, would tend to stay on Earth.

Another crucial invention was the gathering of genetic information into a nucleus within the living cell, as our own chromosomes of heredity are located. It was a first step towards sexual reproduction, which was to accelerate evolution enormously by permitting the constant reshuffling of genes. The oldest known organisms with nuclei in their cells are fossil green algae dating from about 1000 million years ago, found near Alice Springs in Australia.

For life to evolve rapidly in the oceans, it remained necessary to accumulate a sufficiency of oxygen. By about —600 MY the oxygen in the atmosphere was probably at a few per cent of its present abundance and animal life began in earnest. At this stage a notable invention occurred in the design of organisms. Where older sediments remain unmangled by subsequent mountain-building, they lie in tidy layers. Quite suddenly, in overlying layers, the appearance changes, where many sediments have been churned up. The burrowing worm had arrived, the ancestor of all of us who have hollow bodies.

In Asiatic river estuaries there survives one of the most successful of all animals, lingula, anchored to the mud in a tongue-shaped shell. It has existed virtually unchanged for 500 million years. Lingula represents a huge army of related humble animals, the brachiopods. To an unbiased observer – a Martian, say – the most typical inhabitants of the Earth might be the brachiopods of the sea and the insects of the land.

Until scarcely earlier than —400 MY, dry land was no place for plant or animal. The reason was the intense ultraviolet rays from the Sun, which would cause fatal

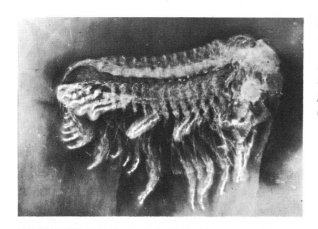

Fossil fishes (right) about 200 million years of age. Reptiles, birds and mammals are all descendants of the bony fishes.

An X-ray photograph (left) of a fossil trilobite, a many-legged sea animal that lived about 400 million years ago, here shown life-size. X-rays make the fossil more lifelike by revealing soft parts that are not normally seen. Below: a very early fossil worm (non-burrowing). It is Spriggina, *from Australia.*

sunburn to any organism daring to push its head out of the protective layers of the water. We have to visualise barren continental rocks for nine-tenths of the Earth's history. Beach-heads could be won only when the oxygen in the atmosphere had risen to about ten per cent of the present amounts.

Then, from the oxygen, a screen of ozone formed high in the atmosphere, sufficient to black out most of the ultraviolet rays. Once land-dwelling plants established themselves they, too, contributed to the Earth's oxygen. But this history suggests that there is no fixed, 'natural' abundance of oxygen gas at the Earth's surface and it may have fluctuated widely from the present amounts to which human life, for example, is adapted.

In any case, the production of our oxygen atmosphere was not straightforward. While it is true that a green plant uses the energy of sunlight to convert carbon dioxide and water into carbohydrate and oxygen, there is a big catch. The process is completely reversible and usually is reversed. When the plant decomposes, or is eaten, the carbohydrate reacts with oxygen to produce carbon dioxide and water. There would have been no net gain in oxygen had this reverse process always occurred. The oxygen now in the air persists because some organic remains were buried out of reach of oxygen before they could decompose. Those were the remains that, in favourable conditions, made coal and oil. Most of the carbonaceous material, whose perpetual entombment is our defence against suffocation, is dispersed in economically uninteresting forms, which is perhaps just as well.

Transgressions and supermonsoons

From time to time, the plants and animals that live in shallow water plaster huge areas of the continents with their corpses. It last happened on a prodigious scale

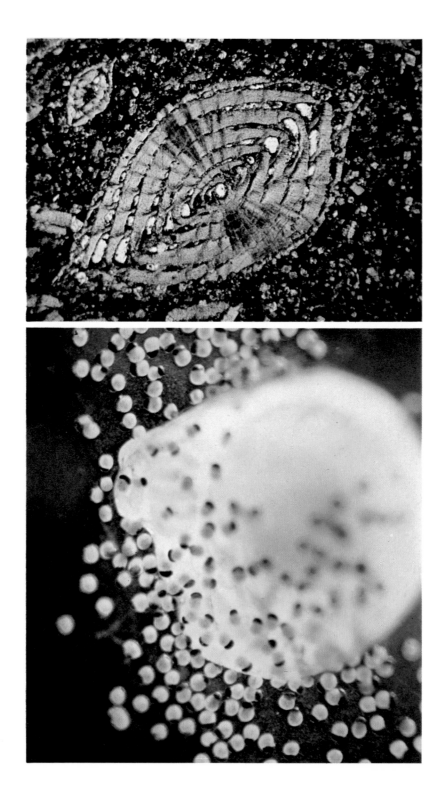

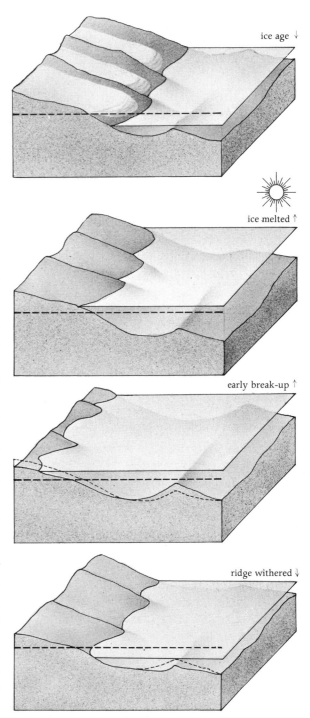

around 90 million years ago, when Europe was coated with chalk and one third of North America was under water. Classical accounts of the Earth's history are, quite rightly, replete with tales of 'transgressions' – of incursions of sea water that flooded large low-lying areas of continents for long periods. The rafts were swamped.

Successive layers of rock often testify to the gradual advance of the sea, as the rocks change from the coarse sands of a beach to the shales made from deep-water mud. The sequence runs backwards as the sea retreats; in between, there may be rocks missing that were worn away while above water. Locally, the wetting and drying may have been due to a crumpling of the land in continental collisions, but there are also world-wide events, with simultaneous flooding of all the continents.

Some changes in sea level occur when large quantities of water freeze in an ice age, or melt again. At present the Earth is in a middling condition with respect to ice. If the prevailing ice melted, the sea level would rise 200 feet; during the most severe of the recent ice ages it was 700 feet lower than it is now. Ice ages are infrequent and brief. Larger and longer-lasting transgressions and regressions, which formerly seemed as unaccountable as Noah's Flood, now have a plausible explanation. It comes from a link between plate movements and sea level.

The mid-ocean ridges, formed where two plates of the Earth's outer shell are moving apart, are enormous structures and they take up a lot of room under water. If all the present mid-ocean ridges vanished, the sea level would drop by well over 1000 feet. The ridges grow at the expense of the adjacent blocks of the Earth's upper layers; as a result, when continents move apart on either side of a ridge, their margins subside, thereby displacing more water. The sea level therefore tends

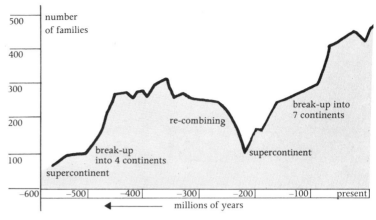

The changing fortunes of marine animals living on continental shelves, indicated by the number of classified families, can be related to the scattering and reunion of the continents. (After J. M. Valentine and E. M. Moores.)

to be high at the early stages of continental break-up. When the continents are bunched together, or when they have cleared the mid-ocean ridges sufficiently to spring up, back to a normal level, the sea level is low.

Very large continents existed during two well-known regressions of the sea: around —700 MY and again around —225 MY. When the respective super-continents broke up and new mid-ocean ridges grew along the cracks, the ridges and the subsiding continents raised the sea level again. The continents were much flooded during busy periods of the two break-ups, at —420 MY and —100 MY. The latter was the period when microscopic sea life laid down the chalk of Europe.

One of the great evolutionary massacres, long known to fossil-gatherers, occurred at —225 MY – the so-called 'Permo-Triassic catastrophe'. It affected marine life in particular and killed off many kinds of animals that had flourished for long periods. According to James Valentine, the reason was that Asia collided with Europe.

Valentine is a theorist of biological evolution in the University of California at Davis, who has a thought-provoking account of the effects of plate movements on life. In particular he speaks of marine animals that live on the edges of continents. In essence, Valentine declares that life was hard whenever the continents were massed together, and comparatively easy when they were dispersed. With his geological colleague, Eldridge Moores, he has related the chequered fortunes of the animals of the continental shelf to the assembly and break-up of supercontinents. The better to follow their reasoning, let us see some clues offered by the present state of the planet.

On the Earth today there is a greater diversity of species of plants and animals than ever before. The continents are well scattered, providing not only separate rafts on which different forms could evolve but also a great variety of habitats, strung out from the Equator to the Poles. In high latitudes, the climate is highly seasonal and from summer to winter animals have to be able to cope with large changes in the amounts of food supplied by plants.

In most of the tropics, the climate is much less variable and all sorts of weirdly specialised marine animals have evolved that take advantage of the steady productivity of the plants. An exceptional tropical area is southern Asia, on the edge of the Earth's largest land-mass. There, too, conditions are highly seasonal, because of the alternating winds of the monsoons. When the monsoons blow offshore, they drag up cold water from the bottom of the sea, which is rich in nutrients; when they blow onshore, the surface water stays warmer, but it is poorer in nutrients.

When most continents were joined together to make a huge supercontinent, Valentine argues, the monsoons were even more marked than in Asia today. In addition, the bringing together of species that were formerly living in isolation presented them all with unfamiliar enemies and rivals. The massing of the continents also lowered the level of water in the oceans and reduced the area of the continental shelf available to the marine animals. All these effects – seasonality, unexpected competition and a falling sea-level – reinforced one another in making conditions very taxing indeed.

Here is ample explanation of the 'Permo-Triassic catastrophe' of —225 MY. It corresponded with the collision between Asia and Europe, along the line of the Ural mountains, which completed the supercontinent of Pangaea.

Continental fusion is a stimulus for evolution, albeit in a harsh kind of way. Diversity diminishes but good new biological tricks may pay off. At about the time of the supercontinent of —700 MY, nature invented the burrowing worm. Its great advantage, in highly seasonal

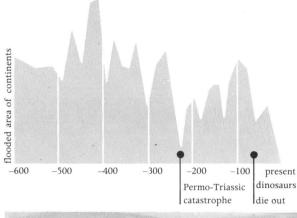

flooded area of continents

-600 -500 -400 -300 -200 -100 present

Permo-Triassic catastrophe

dinosaurs die out

Changes in sea level during the past 600 million years, as indicated by the areas of continents, now dry, that were flooded. The peak 400 million years ago corresponds to nearly 40 per cent flooding. The precise extent of the fall in sea level at the time of the Permo-Triassic catastrophe is uncertain. (Adapted from data of H. and G. Termier.)

Retreats of the sea are recorded (below) as discontinuities in the layers of the Grand Canyon in Arizona. Nearly all the rocks were laid down under water but at the levels indicated by the arrows the sea level was low and the surface here was above water. The big flat surface at the top of the canyon corresponds to the Permo-Triassic catastrophe which extinguished many families of marine organisms.

225 million years ago

about 700 million years ago

The Permo-Triassic catastrophe occurred 225 million years ago, when Asia collided with Europe and the sea level fell.

monsoon conditions, was that it found its food in the stockpile of dead organic remains that had fallen into the mud of the sea floor. For this reason it was immune to the seasonal fluctuations in the growth of living marine plants.

Life is very resilient and evolution proceeds quickly compared with the rates at which continents move. When hard times end – when a supercontinent breaks up or a continent returns to the tropics from a voyage through polar regions – diversity quickly returns as animals exploit the stable tropical climate. Thus evolution is stimulated in a different way. Animals adopt more specialised ways of life and, given higher rates of survival, their powers of reproduction diminish. Once again, many species are in a boom-or-bust situation. But great variety of forms and ways of life brings dividends. For example, when the first plants and animals were moving out of the water on to the land, the chances of success can only have been heightened by the fact that the diversity of aquatic organisms was then at a peak, because of continental fragmentation.

Mammals adrift

When he marshalled the early arguments for continental drift, Alfred Wegener began by pointing out that students of fossils drew conclusions about the history of the planet quite different from those of classical geophysics. Again and again the palaeontologists found themselves forced to imagine land bridges between continents now separated by vast oceans. Those bridges were needed to explain how very similar plants and animals came to be living at the same time on continents that are now disconnected.

The idea of a land bridge is not ridiculous – in previous times you could have walked dry-shod from France to England or from Siberia to Alaska, where now

there are shallow seas. So the theory grew that other continents existed where the oceans now lie, and recently sank. Geophysicists disliked this tale for good reasons, even without present knowledge of the deep ocean floor which puts it out of the question. But the stubborn evidence of the fossils remained; so did the related living species in different continents whose genealogy made no sense without assuming a common ancestry.

Much ingenious thought went into alternative modes of transport across the oceans. Birds could fly, of course, and take seeds and insects with them. Hippopotamuses could swim. But for most of the larger land animals it was necessary to imagine an absurd series of *Kon-Tiki*-like voyages by log to spread the genes around. Take the lystrosaurus, the three-foot reptile whose fossil remains turned up in 1969, just four hundred miles from the South Pole. To reach Antarctica from Africa, where the lystrosaurus was flourishing more than 200 million years ago, the animal would seem to have swum an ocean as wide as the Atlantic. But by the time the polar lystrosaurus was found, the confirmation of continental movements had dispelled all the difficulties. The animal had simply walked from Africa to Antarctica, when those continents abutted.

It is quite hard to believe that the bat, the elephant and the kangaroo are much more nearly related than, say, the shark with the whale. Like the whale itself, they are all mammals. The mammals' rise to prominence coincided with the later phases of the break-up of the last supercontinent and they illustrate better than any other kinds of animals the diversity that comes from living on separate rafts that are drifting apart. Before the break-up, primitive egg-laying mammals had spread throughout the supercontinent. Links between the new continents persisted for various lengths of time, but large-scale flooding of the continents increased the isolation brought about by plate movements.

The northern continents and Africa were the source of the most successful mammals. The pattern is of repeated separations and reunions of the continental areas, which meant that forms could evolve in isolation for a while and then spread and test themselves in new environments. For this reason, it is not always easy to establish what originated where, but the northern continents evidently produced the carnivores (cats, dogs, bears, etc.) as well as other orders of mammals typified by horses, cattle, rats, rabbits, moles, bats, monkeys (primates) and the scaly ant-eater.

On the African raft, elephants, conies and aardvarks originated and also an order of giant mammals now extinct – the embrithopods. When primates found their way to Africa from the north they developed vigorously and much of pre-human evolution seems to have occurred on that continent.

In Australia, the mammalian population did not acquire the placenta – the means whereby most modern mammals carry their young before birth. The primitive egg-laying mammal persisted in the platypus; beyond that are the kangaroos and other marsupials that carry their young in pouches after birth. In South America, too, marsupials flourished but there were 'proper' mammals too in strange forms, many of which are now extinct. At the reunion of North and South America, as recently as —2 MY, more successful species descended from the north on the mammals of South America, and overwhelmed them.

Modern man appeared fully evolved about 50,000 years ago. Again, no one knows exactly where, because his predecessors had moved far afield from Africa. The climatic upheavals of the ice ages, with their demands on resourcefulness, may have helped first to force human evolution and then to 'freeze' it in the most successful type. Ice remains as the variable factor

that bears upon human life on the shortest time-scale – reckoned in decades rather than in the millions of years that are typical of most geological processes.

Archives in the ice

During its history the Earth has usually been a warm planet almost free from ice. The present situation, with Antarctica and Greenland burdened by huge thicknesses of ice, is unusual. During the past two million years, the north polar ice has erupted several times over the surrounding lands. The landscapes of northern Europe and North America owe much to the recent scouring and polishing action of ice sheets and glaciers. At the same time, rainfall has been greater than normal in the rest of the world.

The ice retreated to its present lairs only 6000 years ago. There is no reason to suppose the series of ice ages has finished; rather, we are in one of the 'interglacial' periods which, on past evidence, last for 100,000 years or more. Fluctuations still occur in world-wide climate and the 1970s find us in the midst of a significant cooling of the Earth.

The Greenland icecap is a refrigerated storehouse of snowstorms of the past 100,000 years, neatly stacked on top of one another. Go down, for instance, 350 feet and you will find the winter that drove Napoleon from Moscow. The US Army's cold-region specialists have succeeded in drilling right through the Greenland icecap to the underlying rock. At Camp Century in northern Greenland, where the drilling was done in 1966, the thickness of the ice was 4500 feet. Samples of the ice so recovered went to Copenhagen University, where Willi Dansgaard and his fellow physicists used them to study how the Earth's climate has changed.

The way to find out past temperatures from ice, or from other undisturbed deposits, is not unlike the

Climate during the past 100,000 years shown by the oxygen composition in layers of the Greenland ice. Note the quite sudden onset and cessation of the last ice age, and the warm peak earlier in the present century. (After W. Dansgaard.) The Alaskan glacier, below, suggests what the ice ages were like in wide areas of the northern hemisphere.

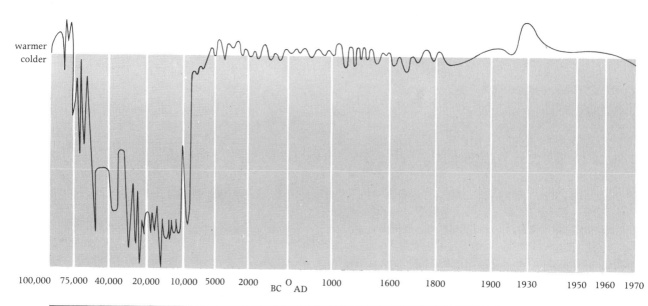

warmer
colder

100,000 75,000 40,000 20,000 10,000 5000 2000 BC O AD 1000 1600 1800 1900 1930 1950 1960 1970

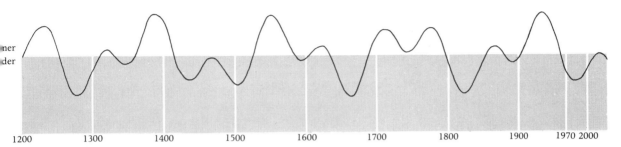

mer
der

| 1200 | 1300 | 1400 | 1500 | 1600 | 1700 | 1800 | 1900 | 1970 2000 |

method used for dating rocks (see page 92). That is to say, it depends on measuring, in each sample, the relative numbers of atoms of the same chemical element but with different atomic weights. In the case of ice, the element used is oxygen, which contains, besides the common atoms (oxygen-16), small amounts of the heavier oxygen-18. Ocean-water molecules containing the heavier oxygen atoms are slower to evaporate than is ordinary water, and especially slow in cold weather. So when that evaporated water falls as snow on Greenland, the amount of heavy oxygen it contains is a clue to the temperature of the ocean; the less heavy oxygen there is, the lower the temperature. In Dansgaard's laboratory in Copenhagen, the instrument which measures the relative amounts of the two kinds of oxygen atoms in each sample of Greenland ice is repeatedly checked against a standard sample of water. Dansgaard calculates the age of each sample from the depth at which it was recovered, taking account of the weight of overlying ice that flattens the deep layers.

In 1969–70, Dansgaard and his colleagues announced their conclusions from the study of the Camp Century ice 'cores'. In the first place they were able to confirm the climatic history of the last ice age and its aftermath that had been deduced earlier by other methods, including atomic measurements in sea-bed sediments as well as geological and archaeological research on land. A parallel investigation of ice, this time from Antarctica, done by Samuel Epstein at the California Institute of Technology, showed that major climatic events in the southern hemisphere closely matched those in the north.

The periods of cold and warmth now determined by all these techniques cast a sidelight on human history. For example, the invention of agriculture 9000–10,000 years ago corresponds closely with the beginning of the end of the last ice age, when the Middle East would have become drier and hunting more difficult. Among

the fluctuations since then, a general warming around 1000 AD created the conditions in which the Vikings of the north could not only harass Europe but also travel to Greenland and America. Conditions worsened in later centuries and during 'little ice ages' Iceland was at times completely ensnared in sea ice. In the 1930s the climate was exceptionally warm. Since then the temperature has been falling again and the Icelanders are again watching anxiously the encroachments of the sea ice.

Not content with confirming the past changes of climate Dansgaard offers a forecast for the future. He bases it on repeating cycles of change that he thinks fit closely with the ice-records of the past – in fact the combined effects of two dominant cycles, one giving peaks of warmth every 78 years, and the other peaks every 181 years. Each of these cycles, Dansgaard believes, represents a regular variation in the Sun which affects its output of energy. Therefore he ventures to let the cycles run on a little way into the decades ahead.

Here is Dansgaard's ultra-long-range weather forecast: the climate will continue to grow colder during the 1970s and early 1980s; then it will become gradually warmer again so that by 2015 we shall be back to where we were in 1960 – no better; and after that it will start becoming colder again. In short, the outlook for the next fifty years is decidedly chilly.

Other investigators of the ice and of climate are cautious about accepting Dansgaard's cycles, and sceptical about making a forecast from them. One possibility is that the cycles are spurious, a kind of accidental trick of numbers like the 'fact' that every third or fourth President of the USA dies in office. But even if the cycles are real and are due, as Dansgaard believes, to a regular variation in the Sun's output of energy, their predictive value is made more uncertain by possible human influences on the climates.

The northern areas covered by ice during the present series of ice ages, and three possible causes. (a) Continental movements created a confined sea at the North Pole. (b) The collision of India and Asia threw up the Himalayas and cooled Asia. (c) The closing of North and South America reinforced the warm current across the North Atlantic. (Ice data after R. F. Flint.)

There are ample grounds, nevertheless, for pursuing this line of inquiry. One reason is entirely practical. During the warmer decades of the mid twentieth century, the cod moved north into Greenland waters and the Danish government encouraged Greenlanders to develop the cod-fishing industry. During the late 1960s, the number of fishing boats greatly increased, but the total catch fell markedly – the cod were moving off again because of the worsening climate. For the fishermen, indeed for anyone working in the Arctic (including the oil companies), the climate and its trends are of more than academic interest. The great need is for checking and extending the sources of information.

Meanwhile, Dansgaard's forecast of climate stands as a provocation to further research in climatology that may support it or prove it inadequate. Patterns of drought and flood around the world will not be easily predicted. H. H. Lamb of the British Meteorological Office, who is one of the most experienced investigators of past climates, has warned: 'The dangers are that climatic forecasting may be rashly undertaken and prematurely combined with attempts to modify or control the climatic trend.'

Why ice ages?

Although the icy times we live in are peculiar, they are not unique. There have been ice ages before the present series. Earlier (page 68) we found the evidence of the work of glaciers in the Sahara, around 450 million years ago, being used to show the great width of a former Atlantic Ocean. Since then there have been other ice ages affecting various continents. The most notable was an extensive one at —280 MY, affecting most of Africa south of the Sahara as well as India, Argentina and southern Australia. This pattern of glaciation, discovered a hundred years ago, made no

sense on the familiar globe and it provided Wegener with one of his strongest arguments for saying that these southern continents were joined together. Going farther back in time, about a dozen episodes of icing occurred in the period —2500 MY to —600MY.

Gordon Robin of the Scott Polar Research Institute at Cambridge remarks wryly of the attempts to account for the ice ages: 'It seems fortunate that we are limited at present to around 60 proposed explanations.' For the demise of the dinosaurs the suggestions must be just about as numerous. I do not intend to mention them all.

Explanations for the onset of ice ages have to allow that Antarctica has been stationed over the South Pole since long before the present series began. Its presence there is presumably necessary for the ice ages, but not sufficient. So we have to look for reasons for a cooling and spread of ice in the north. An awkward point is that excessive cooling would prevent the ice spreading, by cutting down evaporation, and therefore snowfall, in the sub-arctic regions. Hence comes one suggestion for a cause of the ice ages: that the closure of the isthmus of Panama between North and South America, at about —2 MY, deflected all the warm water from the Gulf of Mexico into the Gulf Stream, the warm current that runs north-west across the Atlantic. Its effect would have been to increase evaporation and snowfall in high latitudes. Certainly, changes in the patterns of ocean currents must always be among the more important consequences of continental movements.

Further continental-drift suggestions for the ice ages are less paradoxical. One is that the collision of India with Asia created the Himalayas and an unusual pool of cold at low latitudes. Another is that northward motions of the northern continents hemmed in the Arctic Ocean, isolating it more and more from the warmth of other oceans and so allowing it to freeze

readily. Combine that with the intensification of the Gulf Stream and it might seem that we already have ample reason for the ice ages. At least it has set the stage.

The enthusiast for plate movements may be tempted to look to extensive volcanic eruptions, blanketing the upper atmosphere with dust, as a cause of the ice ages, but he should be quickly warned off. Although H. H. Lamb, who has closely studied the effects of volcanoes on climate, believes that big events like the explosion of Krakatao in 1883 cause a notable lowering of average temperatures in the first following year, he denies that volcanoes could have promoted the ice ages. There is no evidence of exceptional volcanic activity at the onset of the present ice ages.

A change in the gases of the atmosphere is a possibility that needs to be taken seriously. An increase of carbon dioxide would help to trap some of the heat of the Earth that is at present radiated into space. If its effect were to increase evaporation from the polar regions it could have triggered the ice ages. But to increase carbon dioxide greatly, in natural conditions, is not easy. In the case of the disappearing dinosaurs, where longer time-scales are permitted, a fall in the air's supply of oxygen cannot be ruled out.

A choice of catastrophes

For the ice ages and also for the death of the dinosaurs there is a class of explanations that envisage a catastrophe from outer space. In accounting for ice ages, the idea that the Sun blinked has not been much favoured; nor has the suggestion that a cloud of meteoritic dust enveloped the planet. Among the astronomical disasters invoked for killing off the dinosaurs, stellar explosions are perhaps the favourite at present.

Nearby stars must sometimes explode, in the events known to astronomers as novae and supernovae. The effect will be to spray the Earth with abnormal doses of cosmic rays. According to one estimate, very severe radiation from exploding stars will hit the Earth on average once in 300 million years – quite enough (at 1500 roentgens) to kill all large animals on land. Lesser events (500 roentgens), occurring every 50 million years or so, would kill many large animals and sterilise most of the remainder. Plants and insects are more resistant to the effects of radiation.

Another possible source of intense radiation has emerged from astronomy in the past few years. The galaxy we live in, the Milky Way, is one of the vast collections of stars composing the universe, and it turns out that explosions, major or minor, sometimes occur in the centres of galaxies. A very big explosion would destroy life throughout the galaxy; a minor one could add greatly to the doses of natural radiation. Dutch astronomers have found evidence of a modest explosion in the centre of our galaxy that might have affected the Earth about 30 million years ago.

Although catastrophic events far away in space must have sometimes affected life on Earth, the problem is to find any convincing evidence for them. The same difficulty applies to another astronomical prediction, that the Earth cannot help colliding, from time to time, with comets – great masses of ice and dust that were left over in the making of the solar system. The effect would presumably be like a huge explosion – but where are the marks of such events?

One widely publicised explanation for biological catastrophes seems to be ruled out. The discovery that the Earth's magnetism flips over every so often led to the suggestion that, during a period of field reversal, the Earth would be without the magnetic screen that wards off some of the continual rain of cosmic rays.

The idea was that the reversal would bring a marked

A sense of time. For every day that Homo sapiens *has spent on Earth so far, the dinosaurs had eight years.*

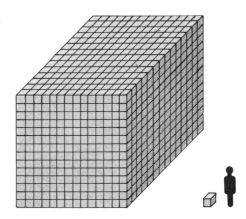

increase in cosmic radiation at the Earth's surface, which would cause lethal mutations in animals exposed to it. But this explanation does not work; most of our protection from cosmic rays is afforded not by the magnetic field which deflects them but by the atmosphere which absorbs them. If the Earth's magnetic field completely disappeared, the radiation dose at the Earth's surface would increase by less than ten per cent – far too little to account for the extinction of whole species of animals. More credible is the influence of the magnetic reversals on the Earth's climate. The effect would be due, not to cosmic-ray particles of very high energy, but to the warming effect of the much more abundant particles of the 'solar wind', which blows continuously from the Sun and is mostly blocked by the Earth's magnetic field. The stronger the magnetic field, the colder the climate. Direct evidence for this connection comes from studies, made at the Lamont Observatory, of magnetism and animal remains in sea-floor samples.

It turns out, in fact, that we should stand the catastrophe question on its head. As well as asking why certain catastrophes have occurred we should also want to know why they have not been more frequent and more varied in kind, given such a catalogue of likely causes. There must be safeguards operating at the planet's surface that tend to prevent disasters from turning into catastrophes – safeguards that we have not yet identified.

Large changes in the amount of oxygen in the air could plainly be quite devastating and have been invoked as causes of catastrophe. It is not at all difficult to envisage either increases or decreases occurring. Any persistent unbalance, however slight, between the rate at which plants produce oxygen and the rate at which the oxygen is consumed would be magnified quite quickly, by geological standards. We should

therefore expect to see effects of wildly erratic variations in the amounts of oxygen in the atmosphere, even to the point where all the oxygen and hence all animal life disappears.

In practice, a balanced supply of oxygen seems to have been uncannily well preserved. A recent investigator of possible mechanisms, Leigh Van Valen of the University of Chicago, admitted defeat: 'A stronger regulator seems desirable but remains to be found.'

One might parody the weirdly varied explanations of catastrophe and suggest that the dinosaurs died out because a piece of the Moon fell off and splashed away half the water in the oceans. A more reliable explanation of the dinosaurs' misfortune must come some day, as our knowledge of the Earth grows. It may be appropriately mundane. The surest evidence of a large-scale change covering the crucial period is of a world-wide regression of the sea occurring between −90 MY and −60 MY, bringing it close to its present level. The arrested growth of the North Atlantic and the upspring of the continental margins of the South Atlantic are possible in terms of plate movements. The withdrawal of the water and the indirect effects on climate and plant growth could well have altered living conditions greatly to the disadvantage of the dinosaurs. Their death was also preceded by enormous deposition of chalk, which may have robbed the atmosphere of much of its carbon dioxide, thus affecting climate and plant growth in another way.

The causes of catastrophe have to be sought, and not merely to satisfy morbid curiosity. It is now a matter of practical urgency. Human activities are altering the Earth in unprecedented ways and with unpredictable effects. Presumably we wish neither to provoke a premature eruption of the ice nor to follow the dinosaurs into the fossil graves.

Unearthing a dinosaur in Alberta, Canada, 1912. This is Corythosaurus, *a web-footed, duck-billed animal – just one of the extraordinarily diverse reptiles that ruled the Earth until 70 million years ago.*

Mining iron ore at Mount Tom Price. The very old hills of Western Australia contain an enormous quantity of iron.

Chapter 6   The New Geological Force

The human species is beginning to rival even the ponderous processes that make and break the Earth's crust and move the continents. Some of the earthquake zones may soon come under our control. But will we learn to match our demands on resources to the rate at which nature renews them?

Shatter and shovel, shatter and shovel . . . the blasting, digging and carrying away of a mountain calls for a measured rhythm of work. Each bite of the mechanical shovel removes 25 tons of rock; each truck delivers a hundred tons to the crushers; each train sets off with 15,000 tons of material for the journey to the sea. Mount Tom Price consists mainly of rich iron ore, so the Australians are taking the mountain to pieces and shipping it to Japan, where it will make super-tankers and washing machines.

The Hamersley iron company's operation at Mount Tom Price is only one of several huge mines recently opened in Australia. The mountain was unknown and unnamed until 1962; by 1982 or thereabouts only the unwanted northern lineament of the mountain will remain, when all that is worth taking has gone and the miners have moved to other, nearby hills.

The red hills of the Hamersley Range of Western Australia would all be judged extremely valuable if they occurred in Pennsylvania or the Ruhr. They were formed from mud rich in iron that settled in a shallow sea, when the Earth was just half its present age. They hold more than a million million tons of potentially useful iron ore, enough to supply the whole world for a thousand years at the present rate of consumption. That the Australian authorities formerly banned the export of iron now seems laughable. Shortage of iron is not going to be an immediate problem for mankind, least of all for the Australians. At present, though, most of the Australian iron ore is pretty valueless, lying as it does in a remote semi-desert.

Only where the mountains are a shiny purple, as at Mount Tom Price, is it worth man's while to set up townships in the desert to mine the iron. The change in colour occurred long ago when, here and there, water dissolved away the mud and left almost pure iron ore, so that two-thirds of the rock is iron.

The mine is opencast, so you can see the ancient mountain being torn apart. No army of miners scrapes away at the rock like the slave gangs of old. A couple of hundred men – Australians, New Zealanders, Englishmen, Germans – use explosives and machinery to clear 30,000 tons in a shift. Watching the shovels biting into the shattered rock was for me as awesome as looking into a volcano. It showed how quickly industrial man is able to alter his planet.

According to legend, King Canute let his feet get wet, to show his sycophants he had no authority over the tides. For a thousand years since then, any idea that human beings could influence significantly the events shaping the face of the Earth has been presumptuous. Knowing more, as we now do, about how these events occur makes human intervention seem, at first sight, even more far-fetched. How can we, who can only scratch the outermost rind of the Earth, influence processes that originate hundreds of miles beneath our feet? What latter-day Canute will command Africa to halt in its tracks before it again slams violently into Europe?

A second glance, though, shows that human beings are not entirely the passive victims of a restless planet. The growth of oceans and the tracks of continents will remain for a long time, possibly for ever, beyond our influence, but these major changes are slow. Californians can view with detached curiosity the prediction from plate movements that Los Angeles will eventually be a part of Alaska. No coast guard in Maine or Donegal need watch for another disappearance of the Atlantic Ocean. But when we consider the face of the Earth as we know it, and the geological and climatic events that affect contemporary life, human powers no longer look puny.

For the first time a species has appeared that consciously alters the face of the Earth. Corals make reefs, beavers build dams, rabbits dig tunnels, sea birds make mini-mountains of guano from their droppings, locusts destroy foliage over large areas – but none of these species is moved by any romantic plan or acts with anything more powerful than its own muscles. The effects of other life, described in the preceding chapter, are geologically significant on long time-scales – thousands or millions of years rather than a human lifetime. The mental powers and technical dexterity of man are much faster and more forceful in their effects.

Unhappily, many people in many parts of the world remain the victims as 'diseased nature oftentimes breaks forth in strange eruptions'. To natural disasters, human clumsiness now adds unnatural earthquakes. Yet, as we shall see in the first part of this chapter, one of the early benefits of the new understanding that earthquakes result from plate motions will be a start in moderating their violence.

Natural disasters

The Japanese have put considerable effort into earthquake prediction. In a room at the Earthquake Research Institute of Tokyo University, signals from instruments scattered around the countryside are brought together to allow a constant watch on the activity of the ground and to provide the information for an earthquake prediction service. The Institute was able successfully to issue warnings for periods of potential danger during a long series of earthquakes at Matsushiro in 1965–7. Before the major earthquake at Nigata in 1964, a tilting of the ground had been detected.

Another policy for the earthquake researchers is to try to recognise, among other minor shocks, the so-called foreshocks that hindsight shows to have preceded severe earthquakes. And the would-be forecasters are making promising observations of electro-magnetic changes in the rocks, as a possible way of keeping track of an unusual build-up of strain. Nevertheless, the state of the art in earthquake prediction was summed up for me by the flashlight that a senior earthquake specialist in Tokyo always carries in his pocket in case of unexpected disaster.

On Hawaii, the scientists at the Volcano Observatory were more confident about the predictability of volcanic eruptions. An eruption usually gives some warning, if there are experts or at least some sharp-eyed citizens keeping an eye on the subterranean weather. Before an eruption of Hawaii's active Kilauea volcano, instruments called tiltmeters record a swelling of the mountain and characteristically rhythmic ground tremors are picked up by seismographs. Other instruments measure stretching or movement of the ground. The chief snag is that threatening signs can turn out to be false warnings, when the volcano relaxes without erupting. Everyone knows that Kilauea is going to erupt at some time, just as everyone knows that Tokyo is going to have a bad earthquake at some time. The problem is to give warnings specific and reliable enough for the public to take them seriously.

All coasts near to regions of earthquakes or explosive volcanoes are vulnerable to tsunamis, often called 'tidal waves' although they have nothing to do with the tides. Tsunamis are huge waves created when an earthquake, a volcano or a slumping of material on the sea floor disturbs the ocean. When the volcano Krakatao blew up in 1883, the waves killed 36,000 people along the coasts of nearby Java and Sumatra. Because tsunamis dash across the Pacific at several hundred miles an hour, an international warning centre operates in Hawaii; but little warning is possible close to the source of the tsunami.

If ever a volcano was set off by human agency, the secret was well kept. The closest I can come to finding

Crump Well, a man-made geyser in Oregon. It has spouted rhythmically for more than 16 years, since it was accidentally started during a drilling operation in search of natural steam.

anything of that kind is a man-made geyser in Oregon, USA. It began spouting in 1955, from a hole that was bored in the course of a search for geothermal power. The hole penetrated a layer of sand containing hot water. Although Crump Well, as it is called, is not an outstanding geyser, its rhythmic behaviour is impressive – once every nine hours steam pressure builds up and shoots a jet of hot water a hundred feet into the air.

Control of the course of volcanic eruptions is no longer entirely far-fetched, at least for predictable and not too violent volcanoes. Where plugs tend to form in the throat of the volcano, allowing pressure to build up dangerously, explosives might keep them open – although no one would be well advised to use nuclear bombs against a volcano. If techniques improve for mapping the arteries within a volcano, it should be possible, in some cases, to make openings in the mountain at chosen points and lead the lava away along pre-selected paths.

Man-made earthquakes

In 1935 the Colorado River was dammed, creating Lake Mead. As the lake was filled, and during the following ten years, 6000 minor earthquakes occurred in what was previously an earthquake-free zone, as the underlying rocks took the load of ten cubic miles of water. In 1963–6, similar but more violent events followed the construction of the Kariba Dam in Africa. In India in 1967 a strong earthquake, probably caused by the filling of the Koyna Dam, killed 177 people living nearby and also cracked the dam itself.

Nuclear explosions, too, can cause earthquakes, at least locally. That is to say, the immediate shock of the explosion, which has a seismographic signature different from an earthquake's, may be followed seconds or hours later by more natural earth movements. For example, an underground test called Boxcar involved detonating a 1·2 megaton bomb at the Nevada test site in April 1968. It gave rise to thirty small earthquakes within three days. The centres of the earthquakes lay mainly along a six-mile line running southwest from the site of the explosion, suggesting that an old fault may have been reactivated. Minor tremors continued for several weeks.

Man-made earthquakes entered the realm of geopolitics in 1969, when Canada and Japan protested unavailingly to the United States at the start of a series of tests of powerful nuclear bombs at Amchitka in the Aleutians. The site is right on the northern edge of the Pacific plate which runs to Japan in the west and near to Vancouver in the east. Alaskans remembered the very severe natural earthquake at Anchorage, in 1964, and the other nations had good reason to protest. A year before, Gordon MacDonald, one of President Johnson's science advisers, had speculated about the possible military use of carefully timed explosions further along the rim of the Pacific plate in the China Sea and the Philippine Sea, to knock down San Francisco. Whether an earthquake or explosion can in fact trigger other earthquakes at a great distance is a warmly debated question but the weight of the evidence is against it happening.

A mild earthquake at Denver, the state capital of Colorado, in April 1963 was the first experienced in the area since 1882. For several years thereafter, earthquakes occurred almost daily and one in August 1967 did a little damage in the city. The US Army's Rocky Mountain Arsenal, ten miles from Denver, turned out to be the unintending promoter of the earthquakes.

Nerve gas and insecticides, manufactured at the arsenal, are very toxic and the managers had large quantities of chemically contaminated water to be rid of. For nearly ten years, the waste went into holding ponds on the surface, but escaping material killed crops, livestock and wild life in the area. So the Army made a well two miles deep and began pumping the contaminated water down to where no one imagined it could do any harm. Six weeks later came the first earthquake. Altogether, more than half a million tons of water went down the well before pumping ceased in February 1966.

The well reached down into very old rocks, which were evidently fractured and already under stress, close to their breaking point. The effects of the water under high pressure were to force cracks to grow and to allow the rocks to move more freely over each other. A year after pumping had ceased, the water was still pressing outwards from the base of the well and reactivating some of the longer cracks – which gave rise to Denver's biggest earthquakes in 1967.

The story of the Denver earthquakes gradually emerged from 1965 onwards, in spite of official dissimulation. The frequency of earthquakes turned out to be closely related to the varying rate of injection of water. The excitement of the earthquake experts reached a peak when the earthquakes virtually ceased, a couple of years after the Army stopped using the well. The Denver earthquakes had been switched on by human agency and switched off by human decision.

Across the Rocky Mountains from Denver, at Rangely, Colorado, man-made earthquakes of just the same kind occurred in an oil field, in the 1960s. The Chevron oil company pumped water into the ground to increase the supply of oil, and thereby caused some minor earthquakes at one side of the field. In 1970–1, US Government geologists from the National Center for Earthquake Research took over part of the oil field for the first deliberate experiment in the control of earthquakes. They set up a network of seismic instruments

Pouring slag at an ironworks. Man alters the local composition of the Earth's surface, both by extracting minerals and by generating the by-products of industrial activity.

to record and pinpoint the earthquakes then in progress.

In November 1970, the experimenters began removing the water from the ground. A few months later earthquakes in the vicinity had almost entirely ceased. Then, in April 1971, water was pumped back into the ground and before long earthquakes began again. In this little corner of Colorado, at least, the solid Earth was under human control.

Save San Francisco!

The Earth-tamers are an unpretentious group of scientists going on quietly with their work in the US National Center for Earthquake Research. The Center is at Menlo Park, which is appropriately close to San Francisco. Like many other travellers, I count San Francisco as one of the world's finest cities and grieve that it is threatened with a major earthquake, comparable with the one that assailed it in 1906. The disaster may happen before these words are in print, or it may be delayed another half-century. The longer the delay, the worse the earthquake will be, because strain is building up steadily in the rocks, like a spring under increasing tension. The people of California have been casual about their earthquake hazards and many buildings, including schools and hospitals, straddle the fault lines.

Pessimists talk of 50,000 dead in San Francisco next time, and of damage at 50,000 million dollars. Yet, following their successful experiment in earthquake control at Rangely, the research workers at Menlo Park think they know how, given time and money, they might save San Francisco.

The idea of earthquake control by water injection, demonstrated at Rangely with real earthquakes, is also a subject of laboratory experiments. At Menlo Park, James Byerlee produces miniature earthquakes in rock samples with the aid of a powerful press. The rocks break under the applied pressure and then the surfaces of the fracture continue to grind over one another, sticking and slipping in jerks, which are recorded by detectors attached to the sample. But Byerlee is also able to pump water into the pores of the rock, under high pressure. When he does so, the surfaces of the fracture move more easily, with less sticking; the jerks become more frequent but much weaker. By tests of this kind on rock samples from areas actually affected by earthquakes, Byerlee can advise his colleagues on how water under pressure might affect the motions of the Earth.

The natural motions near San Francisco are awesome enough. The city is caught between two immense millstones, the American plate which stretches from California to Iceland and the Pacific plate which extends from California as far as Japan. The main plate boundary is the San Andreas transform fault, which runs through California and passes out to sea within a few miles of downtown San Francisco. Subsidiary faults zigzag through the region. The Pacific plate sidles northwards past America at $2\frac{1}{2}$ inches a year, on average, and no conceivable human force is going to stop that motion, which originates deep in the Earth. At present, the rocks in the San Francisco area have stuck in an ominous way; they may be as much as 13 feet in arrears, compared with the general motion of the plates that has occurred since the time of the 1906 earthquake.

He would be a rash man who tampered with this situation simply by pumping as much water as possible into the San Andreas fault near San Francisco. That would certainly encourage it to slip and catch up with the Pacific ocean floor, but it might well do so with precisely the major earthquake that is now feared. For a rational scheme, just as important as the new-found power of water pressure to make movements occur

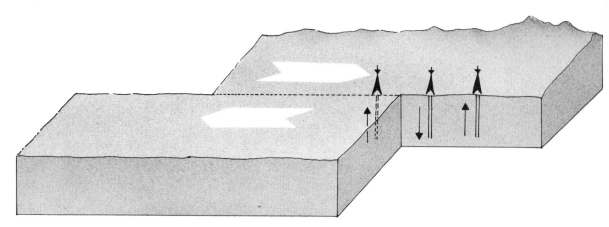

more easily is the opposite effect: that pumping water out of a fault zone can lock it, and discourage movements.

As Barry Raleigh of the Menlo Park laboratory outlines it, the desperate scheme for preventing the next major earthquake at San Francisco involves causing a long series of minor earthquakes. The violence of the earthquakes would be kept within predetermined limits. The first step would be an experiment in a fault similar to the San Andreas fault, but far from centres of population. The Fairweather fault in Alaska is a candidate.

Three holes would be bored along the fault at intervals of about 500 yards, each of them reaching two and a half miles into the Earth. Pumping water out from the two outlying holes would be the next procedure; doing that would tend to lock the fault at those points. Then, when water was pumped into the middle hole, no earthquake that it provoked should propagate along the fault beyond the locks of the two other wells. After the small man-made earthquakes had eliminated the strain in the region, the middle hole, too, would be pumped dry, to lock the fault more securely.

If that pilot experiment were successful in removing the strain from the selected strip without causing any unexpected trouble, the same technique could be applied all along the San Andreas fault, including the region close to San Francisco. The triplets of holes might be spaced at intervals of six miles; any earthquakes occurring spontaneously within the intervening region would be insufficient to cause much damage. So for the California fault lines, 500 boreholes would be needed, at a cost of about one million dollars each – an expensive operation, but modest compared with rebuilding a big city after an earthquake.

Even so, the work could not be done overnight; if a start were made right away, the San Andreas fault might be tamed in about ten years' time. Thereafter, provided the holes themselves were not damaged by the man-made earthquakes, they could be used repeatedly to ease the northward passage of the Pacific plate boundary through the Californian countryside.

Could this lifesaving application of the new understanding of the Earth also work in other regions afflicted by earthquakes? California is unusually fortunate in that the plates seem to be self-lubricating below a depth of about twelve miles. All Californian earthquakes originate fairly close to the surface, and may be controlled with boreholes of normal penetration. In some parts of the world, earthquakes occur much farther below the surface, especially where they are due to descending plates that are being swallowed into the Earth at ocean trenches.

For this reason, many zones could not be freed from earthquakes by the Menlo Park technique; they include Alaska, Japan, the Philippines, Indonesia, Greece, Chile and Peru. It is also beyond the bounds of present technology to tackle earthquakes that do not occur under land. Nevertheless there are a few areas where the situation is broadly similar to that in California. Perhaps the most notable are northern Turkey and Iran, the scenes of great earthquakes that have caused much loss of life. For the areas where control is at present out of the question, the remaining hope is that reliable means of forecasting large earthquakes will become available in the years ahead.

Of reefs and trenches

The present plight of the Earth's coral reefs is perhaps the most dramatic example of how shifts in the ecological 'balance of nature' can have far-reaching geological effects. Nobody knows why the spiny starfish *Acanthaster planci*, formerly quite rare, suddenly multiplied

The strain in a transform fault might be relieved (left) in a series of mild earthquakes, if the fault is 'self-lubricating' at great depth. The fault would be locked at two points (by pumping water out) before causing the earthquakes (by pumping water in). Finally, the whole section would be locked (by pumping the water out again). The San Andreas fault (right), here appearing in the Californian countryside, is a suitable case for treatment by this method.

San Francisco. Rebuilt as a modern city since the 1906 earth-quake, it now awaits another lurch north-westwards by its suburbs on the Pacific shore.

An island protected by coral, in the Great Barrier Reef of Australia. Annihilation of the coral by the starfish now on the rampage would have disastrous consequences for many inhabited islands throughout the Pacific and Indian Oceans.

Starfish attacking coral. The coral on which the spiny animal lies has been killed. There is living coral to the right.

out of control in the mid-1960s in the Indian and Pacific Oceans. It feeds on the soft tissue of living coral and in 1966 it began a massive attack on Australia's Great Barrier Reef, killing the coral over more than a quarter of the length of the reef. The starfish explosion spread across the Pacific, wiping out much of the coral around Guam and other islands and atolls. The starfish was previously kept in check by the coral itself, which ate its larvae.

Richard Chesher of the University of Guam suspects that human blasting and dredging operations started the coral disaster by creating dead patches where the starfish larvae could survive in greater numbers. Now the coral killed by the adults provides safe breeding grounds. As if the starfish were not enough, a quite different enemy assails the coral of the Hawaiian island of Oahu. A filamentous alga forms mats and smothers the coral. The eruption of this plant is almost certainly due to domestic sewage running into the ocean.

The wholesale killing of reef-building coral will, if it continues, have extraordinary consequences. First, the way of life of the Pacific islanders will be destroyed, because it is based on fishing and fish desert dead coral. Then their very islands may be destroyed, as the dead coral is eroded and the reefs which serve as natural breakwaters cease to be effective against storms. Continental margins, too, which are at present protected by coral, may also suffer accelerated erosion.

If, as Chesher thinks possible, we are witnessing the extinction of the reef-building corals of the Pacific, the Earth will soon be a noticeably different planet and, if there are geologists ten million years from now they will see the extinction recorded in the rocks just as geologists of our time can point to the sudden disappearance of important species of the past. So if man is even in part responsible for the death of coral, he may have made an indelible mark on the planet. Other human effects are more diffuse.

The undisturbed polar ice sheets, besides preserving a record of climatic change (p. 125), also serve as chemical archives of the Earth's atmosphere. They sample thousands of years of snowfall. Claire Patterson of the California Institute of Technology, with his colleagues, analysed the historical ice samples brought by the US Army from Greenland and found progressive changes in the amount of lead that the ice contained. Even in the seventh century BC, the amount of lead brought down by the Arctic snow was noticeably greater than the natural level, because of the exploitation of lead by ancient civilisations, especially in the separation of silver. By 1750 AD, the industrial lead escaping to the atmosphere brought the amount in the snow to twenty-five times the natural level. The really steep rise began after lead was added to motor fuels to make them perform better. The recent snows of Greenland contain 500 times more lead than old ice free from man-made contamination.

Other analyses of polar ice, by Minoru Koide and Edward Goldberg of the Scripps Institution, show that almost as much sulphur is now entering the atmosphere from the burning of coal and fuel oil as normally enters it from the decay of organic material and from volcanoes. The sulphur has increased particularly sharply since 1960.

Often there are two ways of reading the figures. Nature puts at least as much sulphur into the air as comes from human industry. If you are a seller or burner of sulphur-rich crude oil, you might argue that you were much less offensive than a volcano. If you are a member of the public suffering damage to lungs, clothing or trees by sulphur dioxide, you are entitled to say that your fellow humans have increased the input of sulphur into the air by a staggering 100 per cent and have brought the equivalents of volcanoes into cities.

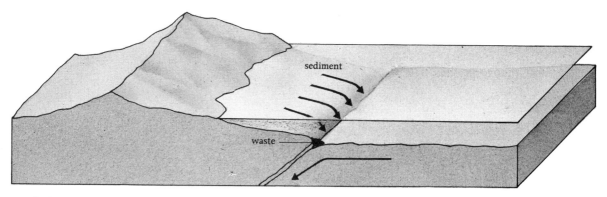

sediment

waste

Similarly, in arguments about whether smoke affects the climate it is not enough to say that the explosion of the Krakatao volcano put more dirt into the air than man has done in his whole career on Earth; one should also note that some climatologists believe that Krakatao did affect the climate. The facts or estimates that earth scientists can now give about natural processes help to put man-made processes into a proper perspective.

Our preference for particular materials in the Earth's crust alters the chemistry of the Earth's surface in ways potentially harmful for life. Materials such as iron, which are commonplace and harmless, give little cause for concern except in the rusty eyesores of scrapyards. Comparatively rare materials which man first concentrates and then lets loose are more alarming – especially, at present, two toxic metals, mercury and lead. Earth scientists should now be able to testify that new mercury and lead come naturally to the surface at so many thousand tons a year, and thus provide a yardstick for the control of such pollution. The Hawaiian volcanoes, for example, feed significant amounts of mercury into the atmosphere. Mercury is a case where the evidence is reassuring, up to a point: the natural flow of mercury into the environment from the ground is much greater than any direct contribution from man's use of mercury. But man may indirectly augment the natural flow, by his digging operations.

There are no natural comparisons for our completely novel materials, including poisons against living organisms, of which the insecticides like DDT are the most notorious. It was shocking but hardly surprising when, a few years ago, measurable amounts of DDT appeared in the penguins of Antarctica. Several governments have acted to ban such materials, but, unless every nation does, the Earth as a whole is still contaminated. The effects on life on land and in the oceans cannot be predicted with precision, but they are likely to include hard times for some species and, as a result, boom times for others. While DDT is still needed to help the sick and hungry in the human population, the dilemma about its use is perplexing.

Michelangelo's sculptures will, no doubt, stay as they are for a long time. But most other materials worked by man soon finish up in the air, in the water or in piles on the ground. Proposals for using deep wells to dispose of radioactive wastes from nuclear power stations, or other noxious liquids, deserve sceptical treatment. Liquids put underground have disconcerting ways of seeping out again. As we have seen, when the US army tried to rid itself of chemical poisons down a borehole, it had unpleasant shocks of another kind.

Every day human beings throw out several million tons of solid waste. Some chemists like to think of great retorts which would renew the materials in the waste. No doubt we shall become more systematic about refilling disused mines and quarries with solid waste. But for those wanting to be simply rid of the stuff for ever, the new geology offers the ultimate waste-disposal unit.

At all the ocean trenches, the Earth is swallowing its own plates, as one dives under the margin of another. Material thus engulfed disappears for a very long time. Even if it re-emerges hundreds of millions of years later it will have been transformed into rock or vapour in the pressure cooker of the Earth's interior. R. C. Bostrom and M. A. Sherif of the University of Washington at Seattle were the first to point out the possibilities of feeding waste material into these 'tectonic sinks'. Means already exist for compressing municipal waste to make it denser than water. In principle, dumping it anywhere in an ocean trench would ensure that the waste eventually disappeared, but in practice the sites should be chosen by careful survey. At some places, the waste would take forty years to sink one metre;

If solid waste from cities were compacted and dumped in a trench where an oceanic plate is descending, it would be eventually carried into the Earth's interior. The best site would be one where abundant sediment coming from a continent buried the waste quickly.

elsewhere, according to Bostrom and Sherif, especially near the mouth of a major river, sediments would bury the material in a matter of months and it would never be seen again.

Rivalling nature

During the past few years, the human world suddenly woke up to see the extent of its despoliation and pollution of the planet. Agricultural, industrial and domestic activities damage and deplete our surroundings and threaten both life and the quality of life. What in 1960 had been matters of specialist concern to sewage engineers, zoo keepers and metal brokers came, by 1970, to be frequently on the lips of presidents and prince consorts. The predictions of the conservationists escalated until we were said to be about to run out of oxygen. New scientific understanding of the Earth comes opportunely when we need more level-headed assessments of our planetary prospects.

As I have already implied, what man does towards altering his planet is less contentious than the *rate* at which he does it. One of the most significant effects of our present conduct may be the increase in carbon dioxide gas in the atmosphere, produced by the burning of coal, oil and natural gas. It must eventually affect the climate, if it is not already doing so. Now, the oceans are a wonderful chemostat for the planet – a regulator that helps to keep the chemical compositions of air and water constant. Huge amounts of carbon dioxide are dissolved in the oceans or lie chemically trapped in lime deposits on the ocean floor. Given time, the oceans could 'mop up' our additions of carbon dioxide, by dissolving it in cold water masses that are travelling downwards. They have been doing so to an impressive degree: the burning of coal and oil has, during the past hundred years, increased the carbon dioxide in the

atmosphere by ten per cent; but that is only half the carbon dioxide that man has added to the atmosphere in this period, showing that the oceans and other natural agents have absorbed the remaining half.

By this little sum, we discover the rate of operation of nature's carbon-dioxide chemostat and learn, fairly exactly, the rate at which man could go on burning fossil fuels without the gas putting the climate at hazard. It would be half the average rate for 1870–1970, or about one-quarter of the present rate. That is the figure to which we shall probably have to cut ourselves back, obtaining the rest of our industrial energy from nuclear power or other non-combustible sources, such as sunlight.

This is not an academic or polemical conclusion: calculations of this kind must surely become the basis of economic policy. If it seems like cruelty to oil companies, one can only say that they are already threatening to put themselves out of business. They, and we who benefit from their products, are squandering in the course of a couple of centuries, resources of oil that the Earth spent hundreds of millions of years in making. Were we to decide on a perfectly reasonable long-term policy of restricting the rate of extraction to the rate of natural production, the human species would have to share about a hundred cars and trucks.

For the first time, we obtain a rough measure of the overall rates at which natural geological processes occur. The Earth's surface continually renews itself. It does so most vigorously at mid-ocean ridges, where the crust cracks open as its plates move apart and volcanoes and upwelling rock fill the gap. Taking all the mid-ocean ridges together, the rate of production of new oceanic crust seems to be about 50,000 million tons a year. To avoid using large numbers tediously, I shall call that Rate 50.

The continents, too, grow by volcanic action and by

mountain-building processes which bring hot rock into the crust from below without its necessarily reaching the surface. The average rate of production of new continental material is considerably less than in the oceans – perhaps Rate 10.

Human beings are now taking materials out of the Earth at Rate 15 – fuel, metal ores, building material and the like. In the USA the rate of mining, quarrying and oil extraction has reached twenty-five tons per person per year. If this were not a peculiar privilege of the United States, with its high living standards, the world total would be nearer Rate 100. These figures do not include the tremendous amount of earth-moving that is done to make embankments and cuttings, to level the ground or to dredge shipping channels. The Soviet Union, in particular, is making a start in faster earth-moving by the use of nuclear bombs, in the first instance to form reservoirs. Human activities are clearly competing with natural processes in the sheer volume of material handled, even without regard to our peculiar selection of particular substances.

The output from a big mine like Mount Tom Price, in tons per year, is greater than the output from many active volcanoes. Each year, the Mississippi, one of the world's great rivers, delivers silt to the Gulf of Mexico at Rate 0·5; each year, man takes oil away from the Gulf area at Rate 0·3. The comparable quantities of silt that the River Nile brought to Egypt and the Mediterranean are now largely arrested by the Aswan Dam. The Rhine and other rivers spent several million years building the Netherlands with their silt; by draining the Zuider Zee, the Dutch are extending their territory by six per cent in the course of half a century.

These references to rivers lead us to what may be the most insidious and remarkable effect of human activity. Partly as a result of our farming and land-clearing activities, the rate at which the continents are denuded by erosion has recently doubled. Estimates of pre-human weathering of the continents suggest that the natural rate of erosion proceeded at Rate 9 or 10; this figure is close to the annual formation of new continental material already mentioned. Measurements of present erosion, demonstrable in the material carried by rivers, indicate Rate 20. The ground is, as it were, being washed away from under our feet. Hundreds of millions of years would be needed to level the land, even if that rate persisted. It is the quality rather than the quantity of the loss that is disturbing, because much of it represents soil that is not quickly renewed.

The lesson of copper

The knowledge that alternative sources of energy exist allows us to shrug off the fact that the richer oil fields will be exhausted in the foreseeable future. The same cannot be said so confidently for some other materials in short supply, such as copper and zinc, or even for commoner materials like iron and aluminium, the richer and more accessible deposits of which are being rapidly exhausted.

As their local supplies of minerals dwindle, economically powerful nations look farther afield. The gold-laden galleons of the Spanish Main symbolise one of the incentives of the colonial era. But the acquisition of treasure from the developing countries has actually increased since they achieved political independence and, of course, they are glad of the immediate revenue from exports of ore and oil. Harrison Brown, a geochemist at the California Institute of .Technology who is also foreign secretary of the US National Academy of Sciences, fears that the poorer nations of the Earth will fail to develop their own industrial abilities before the high-grade ores have gone which make industrialisation relatively easy.

For the species as a whole, Brown holds out the comfort of last resort. Man need never abandon industry altogether, because he can ultimately extract his metals from ordinary rocks. All the elements he needs are present in granite together with the nuclear fuels, uranium and thorium, which could supply the energy for extracting the other materials.

'Such a way of life would create new problems,' Brown comments, 'because under those circumstances man would become a geologic force transcending by orders of magnitude his present effect on the Earth. . . . The world would be quite different from the present one, but there is no reason a priori why it should necessarily be unpleasant.'

Unless man changes his industrial habits he may, in the end, be reduced to grinding up Mont Blanc and the Sierra Nevada. The necessary rate of release of heat, far surpassing our present energy supplies, will have considerable ecological and climatic effects and may constitute an 'a priori' deterrent. The scale of operations defies imagination. For example, to sustain our present meagre supplies of copper, using ordinary granite, would require the quarrying and processing of 50,000 million tons of rock a year (Rate 50). A better policy would be to try to match our consumption to the natural rates of production of new ore.

Until very recently, the chemical events at any one place in the planet seemed isolated and arbitrary acts of nature. One of the most instructive aspects of the new geology is its picture of a sequence of chemical events occurring to the same parcel of material at different places, from its formation at the mid-ocean ridge, during its passage across the ocean floor, to its reprocessing at the ocean trench – giving, in other words, a clearer genealogy of minerals than ever before.

The rates of injection of virgin elements into the Earth's surface from mid-ocean ridges are discoverable and, as already mentioned, they should help to give a yardstick for man-made pollution. They will also provide a rough guide to the scale of human mining operations that would not be out of keeping with nature. There are many qualifying factors: interactions, via the sea water, with pre-existing materials; the loss of metals down ocean trenches; the elaborate natural processes of concentrating rich ores; effects of continental collisions and the opening of new oceans; and so on. Nevertheless, we can acquire from the mid-ocean ridges an impression of the slow speed with which the Earth's chemical operations proceed.

Copper provides a suitable example of the kind of calculation that should now be attempted. In electrical machinery, the only satisfactory substitute for copper is silver, and copper starvation will soon become serious. Known deposits are likely to be exhausted early in the twenty-first century. Nothing short of that exhaustion is likely to deter the miners. In any case if living standards are to rise in the developing countries, a huge increase in the inventory of copper products will be necessary.

Many commentators imagine, wrongly, that one needs only to notch up technology a little to extract vast quantities of metal from low-grade ores. That may be true for some very common metals, like iron and aluminium, but it is not the case for copper. Low-grade ore, by definition, has little copper in it. In and around known deposits most of the metal is in the high-grade ore. The long-term supply of new copper for industry may be almost nothing except the new ores made by the Earth.

An interesting but grossly misleading coincidence is that the rate at which virgin copper is injected into the Earth's crust just about equals the present rate of man's extraction of copper – five million tons a year. The new natural copper arrives at the mid-ocean ridges in the basalt that is added to the growing oceanic plates and it comprises about one ten-thousandth part of that material. Much of it is diffused in the rock, but some of it forms a thin, metal-rich deposit on top of the lava. When the ocean plate is destroyed at an ocean trench it takes most – say, eighty per cent – of the copper with it.

Perhaps one million tons of copper a year are 'saved' at the trenches and incorporated into the adjoining island arcs or continents. Some of it forms smeltable ore, but much of it remains diffused very sparsely through ordinary rock. Even if advancing technology made it possible, in principle, to extract metals from the leanest ores or ordinary rocks, we really cannot put all our territory through a grinding mill. Before long, it should be possible to discover what fraction of the ocean-floor copper coming ashore is converted to high-grade ore. Meanwhile, one might guess at one per cent – giving a rate of formation of new copper ore that corresponds to 10,000 tons of metal a year. That figure is probably optimistic, because 2700 million years of supposed plate movements should, at such a rate, have produced far more copper ore than exists in known deposits on the planet.

However one juggles with the figures and the guesses, the long-term supply of copper as high-grade ore is very much less than the present rate of extraction of five million tons a year. If we wish to postpone the need to process vast amounts of ordinary rock, as in Harrison Brown's 'fail-safe' policy, we shall have to make do with a trickle of new metal to replace losses and treat the existing copper like gold. Metal exhaustion is bound to have a profound effect on economic procedures; for example, all equipment containing copper, zinc, lead and tin may come to be rented, rather than sold, to make sure of the recovery of scrap.

Grievous difficulties with those four metals are already unavoidable and, the longer the nations delay facing up to them, the worse the predicament will be. All the new discoveries of ore, and the conceivable new technologies for working low-grade deposits, are likely to postpone the crisis by months, not by decades. For instance, copper is an important ingredient of the manganese nodules that litter the ocean floor. Yet even if, with great expense and effort, the manganese nodules were recovered, they would be used up in a very few years to maintain present supplies of copper. Each coming generation will find itself running into difficulties with further metals: tungsten, nickel, manganese and so on.

Human affairs are already well geared to some of the natural cycles and rates of change: to the Earth's daily rotations, to the changing seasons, to the life cycle of the individual, and so on. We are about to be forced to live, industrially, in accordance with the rhythm of plate movements, moderating our demands on the Earth's resources to natural rates at which those resources are renewed. The new geology has come most opportunely to help us make the unpalatable calculations about the true state of our mineral ecology.

Besides using low-grade ores and recycling the existing stocks of materials much more carefully, we must be prepared, even in our own period of worsening shortages, to leave some of the known oil and copper ore in the ground, for future generations. If that last notion is dismissed as absurd or idealistic, the only conclusion can be that our species does not aspire to prosper on this planet for a geologically significant length of time.

A sense of time continuing

I look out of my window in Sussex across a tract of Wealden clay to a wood where men once quarried iron ore. The cannon and shot that harried the Spanish Armada were made of Sussex iron. The iron-rich clay accumulated in the days of the dinosaurs, when Europe and Greenland were still locked together and Sussex lay drowned at what is now the latitude of Algeria. The clay became buried under hundreds of feet of chalk. Much later Africa collided with Europe, not for the first time, and here, in a little tail of the Alps, the chalk and clay buckled upwards. The weather dug out the resulting dome and laid bare the clay.

Yet now I can scarcely see the clay, even in my mind's eye. Instead I glimpse the wood, the flowers and grass, while my attention is unavoidably riveted by a small mammal, a cat waging unsuccessful war on the birds. The trees and the cat are products of the Earth, just as certainly as the clay is.

The cat, which I now suspect of not really trying, is a subtle rearrangement of materials supplied by volcanoes and photosynthesis and the canned catfood company. The atoms composing the cat have done duty in the Swiss Alps, Greenland's glaciers and tropical forests; many of them were gasped out by dinosaurs pursuing pterodactyls or served in the Armada's gunpowder. These atoms are all that the cat has – or that we have. All geology and all life are a continual re-ordering of the same atomic ingredients, battened down by gravity on a small planet of an undistinguished star.

The clay slowly weathers, after its clammy fashion; in due course the material that provides the platform of this little landscape will have slipped elsewhere, by way of the North Sea. Will there still be humans by then, to feed their lazy cats and to fly to Algeria for the sunshine? To the Earth it's a matter of indifference.

Citizens of Pompeii buried in the volcanic ashes from the eruption of Vesuvius in 79 AD. They left the imprints of their dying bodies encrusted in the ashes, from which the forms were recovered by filling the cavities with plaster.

Index